Remote Sensing Techniques and GIS Applications in Earth and Environmental Studies

Remote Sensing Techniques and GIS Applications in Earth and Environmental Studies

Solange Uwera

Delve
PUBLISHING

www.delvepublishing.com

Remote Sensing Techniques and GIS Applications in Earth and Environmental Studies

Solange Uwera

Delve Publishing
224 Shoreacres Road
Burlington, ON L7L 2H2
Canada
www.delvepublishing.com

Email: orders@arclereducation.com

ISBN: 978-1-77407-636-1 (Hardcover)

ABOUT THE AUTHOR

Solange Uwera, M.Sc in Geographical Information Science from The University of Manchester is a young scientist with passion in research, a writer by occasion and she takes pride in educating the community through writing on every occasion. Most recently, her work landed her a position of writing, where she authors books on physical geography and environment. In the universe, "everything is related to everything else and near things are more related than distant things" and this is a powerful concept she has been exploring and writing about. Solange has experience in environmental monitoring, urban green infrastructure, rural development, environmental education and physical geography research.

TABLE OF CONTENTS

GLOSSARY

A

Alluvial - Relating to or derived from alluvium.

Alps - The highest and most extensive mountain range system that lies entirely in Europe, and stretch approximately 1,200 Kilometer across eight Alpine countries

Altimeters - Altimeter or an altitude meter is an instrument used to measure the altitude of an object above a fixed level

Anchovy - A small, common forage fish of the family Engraulidae

Ancillary - Providing necessary support to the primary activities or operation of an organization, system, etc.

Apparent - Clearly visible or understood; obvious.

C

Camouflage - The natural coloring or form of an animal which enables it to blend in with its surroundings

Chlorophyll - Any of several related green pigments found in the mesosomes of cyanobacteria, as well as in the chloroplasts of algae and plants

D

Data - Facts and statistics collected together for reference or analysis

Dissemination - The action or fact of spreading something, especially information, widely.

Doppler Radar - A specialized radar that uses the Doppler effect to produce velocity data about objects at a distance

E

Erratic - Not even or regular in pattern or movement; unpredictable.

Eutrophication - Excessive richness of nutrients in a lake or other body of water, frequently due to run-off from the land, which causes a dense growth of plant life.

F

Falco Sparverius - Smallest and most common falcon in North America

Fossorial - Animal is one adapted to digging which lives primarily, but not solely, underground

Fragmentation - The process or state of breaking or being broken into fragments.

G

Gradient Modelling - A new approach to resource management and forest fire simulation, has been developed to meet these needs in Glacier National Park

H

Haphazard - Lacking any obvious principle of organization.

Hardware - The machines, wiring, and other physical components of a computer or other electronic system

Havoc – Wide-spread destruction.

Heterogeneous - Diverse in character or content.

High Spatial Resolution - Images having higher spatial resolution are composed with a greater number of pixels than those of lower spatial resolution

I

Ibex – A species of wild goat that lives in the mountains of the European Alps

Impinge - Have an effect, especially a negative one.

Infrared Photography - Infrared photography is a look into the invisible world having light beyond 700nm

Interferometry - Interferometry is a family of techniques in which waves, usually electromagnetic waves, are superimposed, causing the phenomenon of interference, which is used to extract information.

Inundation - An overwhelming abundance of people or things.

Invasive - Tending to spread very quickly and undesirably or harmfully.

Invertebrates - Invertebrates are animals that neither possess nor develop a vertebral column, derived from the notochord

L

Lithography - The process of printing from a flat surface treated so as to repel the ink except where it is required for printing

M

Monochromatic - Containing or using only one color.

N

Neo-impressionism - A pictorial technique where color pigments are no longer mixed either on the palette or directly on the canvas, but instead placed as small dots side by side.

O

Off-Nadir - Any point not directly beneath a scanner's detectors, but rather off at an angle

Orthophotos - An orthophoto, orthophotograph or orthoimage is an aerial photograph or satellite imagery geometrically corrected such that the scale is uniform

P

Photoreconnaissance - Military reconnaissance carried out by means of aerial photography

Phytoplankton - The autotrophic components of the plankton community and a key part of oceans, seas and freshwater basin ecosystems

Plankton - organisms drifting in oceans, seas, and bodies of freshwater

Potential Fishing Zone - A proxy to potential shoals of fish aggregation will benefit the fishing community to reduce the time and effort spent in searching the shoals of fish

Primitive - Relating to, denoting, or preserving the character of an early stage in the evolutionary or historical development of something.

R

Radar - A detection system that uses radio waves to determine the range, angle, or velocity of objects

Radio Collaring - A collar with an attached radio transmitter that is put on an animal so that its movements in its natural habitat can be remotely monitored

Reconnaissance - Military observation of a region to locate an enemy or ascertain strategic features.

Reflectance - The measure of the proportion of light or other radiation striking a surface that is reflected off it.

Relief Displacement - Is the shift in an object's image position caused by its elevation above a particular datum

Retrieval - The process of getting something back from somewhere.

Revitalization - The action of imbuing something with new life and vitality.

S

Salmonids - A family of ray-finned fish, the only living family currently placed in the order Salmoniformes

SeaWiFS - A satellite-borne sensor designed to collect global ocean biological data

Sensors - A device, module, machine, or subsystem whose purpose is to detect events or changes in its environment

Software - The programs and other operating information used by a computer

Spatial - Relating to or occupying space

Spearheaded - To lead something such as an attack or a course of action.

Spectral Analysis - Analysis in terms of a spectrum of frequencies or related quantities such as energies, eigenvalues, etc.

Steep - rising or falling sharply; almost perpendicular.

T

Temporal Resolution - Refers to the discrete resolution of a measurement with respect to time

Thermal Sensors - Thermal sensors are found in many laptops and desktop PCs in order to sound an alarm when a certain temperature has been exceeded

Tunable - Able to be adjusted to the correct or uniform pitch.

Turbidity - Turbidity is a measure of the degree to which the water loses its transparency due to the presence of suspended particulates.

U

Users - A user is a person who utilizes a computer or network service

V

Valorization - The valorization or valorization of capital is the increase in the value of capital assets through the application of value-forming labor in production.

W

Wavelength - The distance between identical points in the adjacent cycles of a waveform signal propagated in space or along a wire

Z

Zooplankton - A type of heterotrophic plankton that ranges from microscopic organisms to large species, such as jellyfish

LIST OF FIGURES

LIST OF TABLES

LIST OF ABBREVIATIONS

ACRIMSAT	Active Cavity Radiometer Irradiance Monitor Satellite
ALI	Advanced Land Imager
ALOS	Advanced Land Observation Satellite
AVHRR	Advanced Very High Resolution Radiometer
BMP	Bitmap
CGIS	Canada Geographic Information System
CIR	Colored Infrared Photography
CLI	Canada Land Inventory
CPUE	Catch per Unit Effort
CZCS	Coastal Zone Color Scanner
DBMS	Database Management System
DEM	Digital Elevation Model
DMC	Disaster Monitoring Constellation
DOQQ	Digital Ortho Quarter Quadrangle
EIA	Environment Impact Assessment
EMR	Electromagnetic Radiation
EO-1	Earth Observing-1
EPA	Environmental Protection Agency
ERP	Enterprise Resource Planning
ERS	European Radar Satellite
ESRI	Environmental Systems Research Institute
ESTO	Earth Science Technology Office
ETM+	Enhanced Thematic Mapper
GEMS	Global Education Management Systems
GEOSS	Global Earth Observation System of Systems
GIST	Geographic Information Science and Technology
GOES	Geostationary Operational Environmental Satellite

GPS	Global Positioning Systems
GRACE	Gravity Recovery and Climate Experiment
GRID	Global Resource Information Database
GUI	Graphical User Interface
ICESat	Ice, Cloud,and land Elevation Satellite
IFOV	Instantaneous Field of View
INCOIS	Indian National Centre for Ocean Information Services
IoT	Internet of Things
IRS	Indian Remote Sensing Satellite
IUCN	International Union for Conservation of Nature
JERS	Japanese Earth Resources satellite
JPEG	Joint Photographic Experts Group
LAI	Leaf Area Index
LCD	Liquid Crystal Display
LIDAR	Light Detection and Ranging
LIS	Land Information Systems
LISS	Linear Imaging Self – scanning Sensors
LOD	Levels of Details
LULC	Land Use\Land Cover
MMICs	Monolithic Microwave Integrated Circuits
MODIS	Moderate Resolution Imaging Spectroradiometer
MOPITT	Measurements of Pollution in the Troposphere
MSS	Multi Spectral Scanner
NOAA	National Oceanic and Atmospheric Administration
OCM	Ocean Color Monitoring
OLI	Operational Land Imager
PAN	Panchromatic
PFZ	Potential Fishing Zone
QA	Quality Assurance
QC	Quality Control
SAR	Synthetic aperture radar
SCADA	Supervisory Control And Data Acquisition
SIS	Spatial Information Systems
SOURCE	Solar Radiation and Climate Experiment

SPOT	Systeme Pour l'Observation de Ia Terre
SQL	Structured Query Language
SRTM	Shuttle Radar Technology Mission
SWIR	Short-Wave Infrared
TIFF	Tagged Image Format File
TIN	Triangulated Irregular Network
TIROS	Television and Infrared Observational Satellite
TM	Thematic Mapper
TOMS-EP	Total Ozone Mapping Spectrometer - Earth Probe
TSM	Total Suspended Matter
UAVs	Unmanned Aerial vehicles
UI	User Interface
USGS	United States Geological Survey
USGS	US Geological Survey
USLE	Universal Soil Loss Equation
UTM	Universal Transverse Mercator
WiFS	Wide Field sensor

PREFACE

Earth and its environment have presumed a great deal of importance in recent times as the stress on the environment have been increasing due to the various kinds of natural and human activities. The researchers and the field enthusiasts have shown concerns over the behavior of the earth and the related components.

There are various ways to study the structural and physical changes in the earth and its environment. In this book, the readers have been shown the relevance of the remote sensing techniques and the GIS tools in the study of the earth and its environment. The book lays significant emphasis on the purposes that are served by these tools in the environmental studies and the impact they can have on informing us about the various behavioral patterns of the earth.

The book also aims to enlighten the readers about the various kinds of activities that cause stress on the earth's atmosphere and the environment. The readers are updated about the impact that these tools can have on handling the crisis situations and planning for them in advance.

In the book, the readers are introduced to the concepts of remote sensing techniques and GIS. They are narrated about the history of remote sensing and GIS and the various kinds and applications of remote sensing. The book also explains the various characters of the real remote sensing techniques and the advantages of such techniques in use.

The book then explains the ways in which remote sensing techniques can be applied to environmental studies and how it can be integrated with other data sets to provide better results. Then the book explains the relevance and application of the GIS tools to environmental studies, relating them with effective environmental management and explaining the ways in which it can help improve the environmental assessment.

The readers are further informed about the ways in which remote sensing the GIS techniques can be used to predict and manage disaster conditions in a detailed manner. The book continues to explain the role that is played by these techniques in forecasting the conditions that may be formed in the fishing zone and the ones which may be there in the habitats of the various kinds of wildlife creatures so that necessary actions can be taken in time to improve the efforts.

Then, the book takes the readers through the various techniques pertaining to remote sensing and GIS, which can be useful in the study of environment and earth for various purposes. There is also an explanation of the various kinds of visualizing and modeling techniques used to structure the data in these fields. The readers are further informed about the various challenges that are projected in the use and application of such techniques in the future and given a perspective on the same.

I hope that this book can provide the readers with the insights they are pursuing and is able to deliver on the knowledge that they expect from this book. This book is an attempt to increase the knowledge base on the use and application of remote sensing and GIS tools, which seems to be limited at this point in time and update the recent findings and applications discovered.

I hope that the book is appreciated for the efforts we have tried to put in so that it encourages us ,even more, to work on similar knowledgeable subjects in the future.

Chapter 1

Introduction to Remote Sensing and GIS

CONTENTS

Remote sensing is concerned with the collection of information about any object without even touching the object. On the other hand, the Geographic Information System (GIS) is the system that gathers, manage and analyzes all the geographic data. The chapter consists the topics such as meaning, history, types and applications of remote sensing which are explained in the first half. The mid half consists of the meaning of passive, active remote sensing, the platforms for remote sensing, its characteristics and the advantages and disadvantages. The GIS is explained with its meaning and components. The chapter also discusses the functions, technologies used, and how the data is represented in GIS. The chapter ends with the discussion about the relation of GIS with society.

1.1. INTRODUCTION

Remote sensing is used to analyze or collect information about any object without having any physical contact with the object. It is considered as an Art and Science both as it involves analyzing the data collected through remotely observing an object.

It is used in day to day life through human senses like Vision, Hearing and the sense of smell. The data collected is usually very large in volume which may exceed 1 TB in a day, thus it is clearly big data.

The data collected for Remote sensing could be in any format, be it variations in force distributions or variations in electromagnetic energy distributions. All of them are categorized as big data and investigations are carried out using standard methods in order to drive a conclusion by observing the object remotely.

Remote Sensing could also be termed as "Earth Observations" as it obtains information about objects or areas at the Earth's surface without being in direct contact with them. If we observe our surrounding carefully then we can find out that remote sensing is something that we are following on a day to day basis, for example, reading the newspaper, watching cars driving in front of us and so on.

The remote sensing data could be either passively collected or actively collected wherein passive data collection focuses on the intensity of the data received and actively data collection focuses on data received by devices directly.

Remote sensing basically involves the collection of data from the earth's surface, this involves having an energy source that provides electromagnetic energy that travels through the atmosphere and scatters. Once the energy has been emitted from the target, a sensor is required to collect and record the electromagnetic radiation, which is then processed, measured and interpreted into a pre specified format.

It involves capturing images from the earth surfaces for various wavelength regions to compare and draw conclusions. The images captured could be of energy emitted by the earth's surface or the solar radiation. This is known as active remote sensing, as the energy source is provided by the remote sensing platform and the systems where these observations depend on any kind of external source, are termed as passive remote sensing.

The definition of Remote Sensing (Joseph, 2004) read as: "Remote sensing is the science of making inferences about objects from measurements, made at a distance, without coming into physical contact with the objects under study". Therefore, without having touched the object, with remote sensing method, it can be used to gather information about an object.

In terms of resource management, the extensive applications of remote sensing can be explained and the definition which is used by the United Nations, (as part of the general assembly resolutions A/RES/41/65 , 95th Plenary meeting, 3 December, 1986) is an appropriate definition of remote sensing, according to them "Remote sensing means sensing of the earth 's surface from space by making use of the properties of electromagnetic wave emitted, reflected or diffracted by the sensed objects, for the purpose of improving natural resource management, land use an d the protection of the environment."

The system used for Remote Sensing is set up using various equipment which includes, a source of electromagnetic radiation like Sun, a sensor device which collects and measures the radiations, either emitted or reflected from the earth and a carrier like Rocket or a satellite or it could even be a ground- based stand which transmits these radiations through the atmosphere.

The sensors used as an instrument in the process of Remote sensing could be of two types, the sensor that measures naturally emitted radiations are known as Passive sensors and the sensor that measures radiations of a specific wavelength to illuminate Earth's surface are called Active sensors.

1.1.1. History of Remote Sensing

The history of the remote sensing goes back to the year 1859, when Gaspard Tournachon, a French photographer, took an oblique image of a small village situated near Paris from a balloon. This era recorded the start of the observations of earth and the concept of remote sensing came since then. All over the world, other people also started to follow Tournachon's example.

Figure 1.1. Gaspard Tournachon (known as Nadar), who took an oblique image of a small village from a balloon.

Source: Image by Wikimedia Commons

In the United States, it was the time of civil war in Virginia, the defense positions were detected through the aerial photography that was taken from the balloon (Colwell, 1983). Similarly, the scientific and technical development such as lenses, photography shot up in the period of civil war in the United States. The airborne use of this technique was also speeded up.

Another country that saw fast development was none other than Europe. The airplanes were used on a large scale for photoreconnaissance. This was the time of World War I. The earth observations were taken more from the aircraft than the balloon. It proved to be more reliable and stable.

The civilian use of aerial photography was started between the period of World War I and World War II. Airborne photography was applied to some of the fields such as forestry, geology, agriculture and cartography. Changes resulted in improvements in the traditional cameras, films and other equipment. However, World War II noted the most important developments in photo interpretation and aerial photography.

This phase also witnessed some other developments in this field such as near-infrared photography, radar and thermal sensing. The separate real vegetation was benefitted from the near-infrared photography and the thermal-infrared as it helped it from the camouflage.

However, the first successful airborne imaging radar came to be valuable for nighttime bombing and it was not used for civilian purposes. This system was developed in Great Britain in 1941 and called a 'plan position indicator' by military persons.

After the 1950s, remote sensing systems made an evolving history from the systems developed for the war. There was colored infrared photography (CIR) which came to be valuable for plant sciences.

In 1956, Colwell conducted an experiment to test the color infrared (CIR) photography in the classification and recognition of vegetation types, damaged, stressed vegetation's and for the detection of diseases. There was a significant increase in the progress of the technology of radar thereafter.

1.1.2. Principles of Remote Sensing

There is an emission and reflection of different amounts of energy from different objects. The energy is from different bands of the electromagnetic spectrum.

The properties of both the material and the incident energy (intensity, angle of incidence and wavelength) decide the amount of energy reflected or emitted. The object that reflects or emits the electromagnetic radiation is unique in itself. This factor helps in the detection and discrimination of the surface features or objects.

The reflected or emitted electromagnetic radiation from an object detects the feature of an object and the device is called a "sensor" (Example, cameras and scanner). OK Google which carries the sensor is known as a "platform" (Example, Satellites and aircrafts).

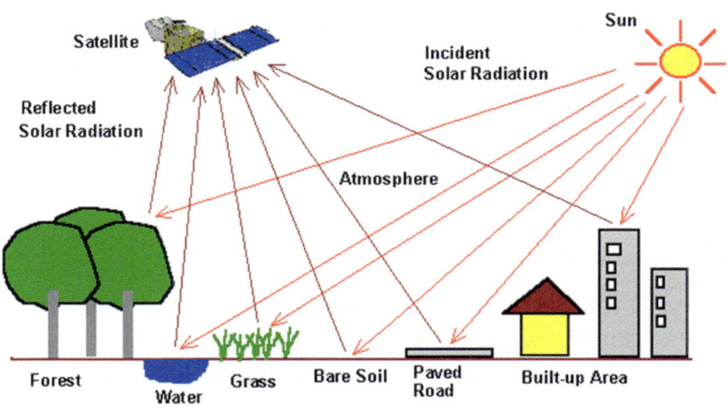

Figure 1.2. The stages in remote sensing.

Source: Image by Principles of Remote Sensing – Centre for Remote Imaging and Sensing

There are majorly 8 stages in remote sensing, these are as follows:

- Emission of electromagnetic radiation - The Sun or the electromagnetic radiation (EMR) source located on the platform
- Transmission of energy from the source to the object
- Absorption and scattering of the EMR while transmission
- Interaction of EMR with the object and subsequent reflection and emission
- Transmission of energy from the object to the sensor
- Recording of energy by the sensor
- Photographic or non-photographic sensors
- Transmission of the recorded information to the ground station
- Processing of the data into digital or hard copy image
- Analysis of data

1.2. TYPES OF REMOTE SENSING

There are different types of remote sensing which have their own feature to suit different analysis. Some of the types of remote sensing are suitable for close scanning while some others are suited for distant scanning. However, the most common among all types of remote sensing is radar imaging.

1.2.1. Radar

The important safety -related remote sensing tasks can be detected through radar imaging. It can accurately detect weather condition and control the air traffic. This can even warn the analysts that bad weather is on its way and whether there are chances of any storm or not. This way the analyst gets ready to cope with the upcoming bad weather.

Figure 1.3. Doppler radar for weather detection.

Source: Image by Wikimedia Commons

In order to collect the meteorological data or to monitor traffic and driving speeds by the enforcement of the law, a common type of radar is used known as Doppler radar. The digital models of elevation can be created by other types of radar.

1.2.2. Lasers

Lasers are the other types of remote sensing. The direction of the ocean currents and the speed of the wind can be measured by the laser altimeters on the satellites. Basically, altimeters, as they have the capability of measuring the bulges of water caused by gravity or seafloor topography, are used for seafloor mapping. It can accurately measure and analyze the varied ocean heights in order to create accurate seafloor maps.

Another form of laser remote sensing is called the LIDAR, i.e., Light Detection and Ranging. LIDAR uses light reflection to measure the distances. This can be best utilized for weapon ranging. The heights of the objects on the ground and chemicals in the atmosphere can also be measured by LIDAR.

1.2.3. Other

There are some other types of remote sensing which include stereographic pairs which are created from the multiple air photos. These are often used to create the 3D featured maps. The air photos obtained through the radiometers, photometers that helps in collecting the emitted radiation from infra-red photos. One such instance is the Landsat program, the air photos that are obtained from the satellites.

1.3. APPLICATIONS OF REMOTE SENSING

The use of remote sensing is diverse and this field is concerned majorly for the processing of the image and its interpretation. The feature of image processing allows the manipulation of the images to create the map. This also helps in saving important information about an area.

The images obtained through remote sensing are interpreted. This can be useful when doing the research for the dangerous or an unreachable area as one can closely monitor and study an area without being physically present there. Remote sensing has been applied in many fields. Few applications of the remote sensing are as follows:

1.3.1. Geology

Mapping is possible in the remote area through remote sensing. It becomes easy for geologists to make the classification of types of rocks of a particular area. The rocks can be studied with regard to its geomorphology. The changes can be traced by this if there are any natural events such as floods and landslides.

1.3.2. Agriculture

While studying the vegetation, remote sensing is very useful. The types of vegetation present in an area can be tracked by the images taken through the help of remote sensing. It can benefit the agriculturists, bio geographers, ecologists and foresters as they can easily detect the potential growth in the vegetation type.

1.3.3. Land-Use Planning

The land use advantage can be studied and regulated through its exposure to wide expanses. Remote sensing can be applied in the land-use planning to detect if there is any usefulness in the land or not. Generally, the data obtained can be applied in city planning and for the modification of the environment.

1.3.4. Geographic Information System Mapping (GIS)

The images taken through the help of remote sensing can be used as input in the case of the raster-based digital elevation models (DEM). The GIS uses the photos taken aerially and later it is digitized into polygons. These are then put into the shape files for mapmaking.

Researchers are the most privileged class of the society as they are using remote sensing due to its varied application. The data can allow users to collect, interpret, and manipulate data from inaccessible or unreachable locations.

1.4. PASSIVE AND ACTIVE REMOTE SENSING

The remote sensing can be classified as passive or active remote sensing. This division is on the basis of the source of electromagnetic energy. The passive remote sensing is the naturally available source of energy. The most important example is the Sun. The solar energy by the sun is used by most of the remote sensing systems and work in a passive mode. This is because there is natural electromagnetic radiation that the sun spreads.

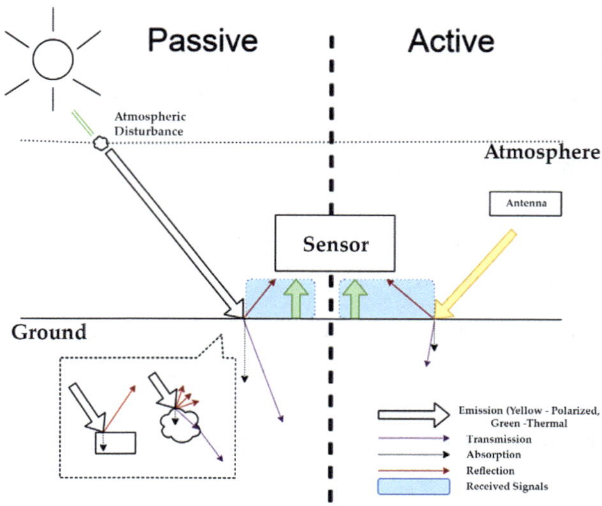

Figure 1.4. Passive and active remote sensing.

Source: Image by Wikipedia

The onboard airborne and space borne platforms record the solar energy which is reflected by the targets at a specific wavelength. This is done using sensors. The object emits some radiation which has a temperature above 0° K (Kelvin). This temperature is approximately proportional to the fourth power of the temperature of the object. Since the earth's ambient temperature is about 300 ° K, it also emits some radiation.

The energy content of the passive sensors is very low and also is not popular. This is the reason that it is not used for measuring the earth's radiance. However, the energy is generated and transferred to the targets from the remote sensing platform in the active remote sensing.

The targets reflect back the energy that is recorded with the use of sensors. The recording is done onboard the remote sensing platform. The active remote sensing is used in most of the microwave remote sensing. In simple language, taking a picture with a camera is passive remote sensing. While taking a picture with a camera having a built-in flash is active remote sensing.

1.5. REMOTE SENSING PLATFORMS

Below are the different types of remote sensing platforms discussed. These types are based on the elevation from the Earth's surface at which these platforms are placed.

1.5.1. Remote sensing at ground level

- Ground- level remote sensors are found very close to the ground
- The ground level remote sensors are basically used for developing and calibrating sensors for different features on the Earth's surface

1.5.2. Aerial remote sensing

- Low altitude aerial remote sensing
- High altitude aerial remote sensing

1.5.3. Space- borne remote sensing

- Space shuttles
- Polar orbiting satellites
- Geo-stationary satellites.

The remote sensing can be either active or passive for each of the above-stated platforms.

1.6. AIRBORNE AND SPACE-BORNE REMOTE SENSING

In the case of airborne remote sensing, the images of the earth's surface are taken by the sensors looking downwards or sideward of the aircraft. This can capture very high spatial resolution images i.e., 20 cm or less. However, this is not suitable for mapping a large area.

Airborne remote sensing has one big disadvantage that it cannot cover large s and hence It has very little coverage. The other negative point is it has a very high cost per unit area of ground coverage.

The airborne remote sensing mission is only one-time operation while the space- borne missions continuously monitor the earth's surface. Remote sensing commonly uses the videography, thermal imagery, LiDAR, digital photography and analog aerial photography.

The space-borne remote sensing makes use of the sensors that are mounted on the space shuttles or the satellites that orbit the earth. Geostationary and Polar orbiting are few to name the remote sensing satellites that help in providing imagery for research and operational applications.

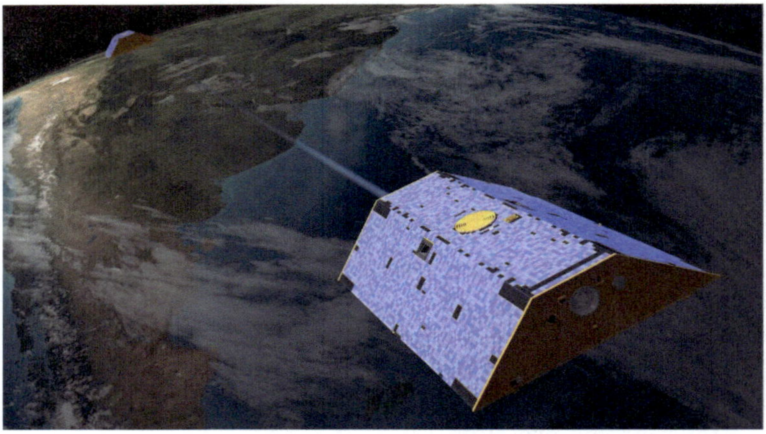

Figure 1.5. Remote sensing through satellites.

Source: Image by JPL NASA

For remote sensing, the most used satellite is the polar orbiting or sun-synchronous satellites while for communication and meteorological purposes, Geostationary or Geosynchronous Satellites are used. It covers up the disadvantage of the airborne remote sensing.

It helps in covering the large area which the airborne remote sensing cannot cover. The per unit area coverage cost is less. It also gives the frequent or continuous coverage of an area of interest, automatic/ semiautomatic computerized processing and analysis.

Aerial photography has one advantage over the space- borne photography that the image captured by the satellite has low resolution. Few examples of the satellites are IKONOS, Landsat satellites, SPOT satellites, Indian remote sensing (IRS) satellites, AQUA and TERRA of NASA and INSAT satellite.

1.7. IDEAL REMOTE SENSING SYSTEM

An ideal remote sensing system consists of the following components:

- A Uniform Energy Source - It provides at a constant rate, high level of output and energy over all the wavelengths.
- A Non-Interfering Atmosphere -It will help in neither modifying the energy transmitted from the source nor emitting or reflecting from the object in any manner.
- A Series of Unique Energy/Matter – The reflection or emission of the signals which occurs from the interaction at the earth's surface are selective with respect to the wavelength. The signals are also unique to each object.
- A super sensor which is highly sensitive with respect to all of the wavelengths. A super sensor would be simple, reliable, précised, cost -effective, and also it needs no power and no space. This sensor produces the data on the absolute brightness (or radiance) from a scene as a function of the wavelength.
- A Real-Time Data Handling System is the system that develops the instance radiance against the wavelength feedback and processes into a format that is interpretable in real -time. The data, which is derived, is distinct in nature with respect to a specific terrain and in this way, it provides insights into its physical chemical-biological state.

Multiple Data Users having knowledge in their respective fields or domains and also with respect to the remote sensing data acquisition and analysis techniques. The information which is gathered will be available to them in less time and also, it is cost -effective in nature.

This information will aid the users with respect to several numbers of the different decision- making processes and also further in executing these sorts of decisions.

1.8. CHARACTERISTICS OF REAL REMOTE SENSING SYSTEMS

The real remote sensing systems are used in general operations, and the utility has several numbers of various shortcomings when they are compared with an ideal system that is explained in the above section of the chapter.

1.8.1. Energy Source

The main source of energy with respect to the real systems are generally non-uniform over several numbers of different wavelengths and also fluctuate with the time as well as space.

This major impact on the passive remote sensing systems. The spectral allocation of the reflected light of the sun fluctuates both temporally as well as spatially. The surface of the Earth also emits energy to various degrees of the effective. A real remote sensing system requires the calibration for source characteristics.

1.8.2. The Atmosphere

The atmosphere alters or changes the spectral distribution along with the strength of the energy that is being received or emitted. The impact of the atmospheric interaction fluctuates with the wavelength associated, sensors used and the sensing application. Calibration is needed in order to remove or compensate these sorts of atmospheric impacts.

1.8.3. The Energy/Matter Interactions at the Earth's Surface

Remote sensing is based on the principle that each and every material reflects or emits energy in a unique, known way. Never the less, spectral signatures may be identical with respect to the several numbers of different forms of materials.

This step makes the process of differentiation a challenging task. And also, in addition to this, the knowledge of most of the energy or matter interactions with respect to the surface of the Earth is either at the elementary level or even entirely unknown.

1.8.4. The Sensor

In real sensors, they have fixed the limit of the spectral sensitivity. In simpler words, they are not sensitive with respect to all wavelengths. In addition to this, the real sensors have limited spatial resolution (effectiveness in recording the spatial details). The selection of the sensor necessitates a trade- off between the spatial resolution as well as the spectral sensitivity.

For instance, never the less, the photographic systems have very good spatial resolution along with the poor sensitivity, non- photographic systems have a poor spatial resolution.

1.8.5. The Data Handling System

The intervention of humans is very significant and vital for processing the sensor data. Never the less, machines are also included in the process of data handling. This makes this idea of real- time data handling almost impossible. The amount of data that is being generated or developed with the help of sensors far exceeds the capacity of the data handling.

1.8.6. The Multiple Data Users

Talking about the success of any remote sensing mission lies on the user who ultimately converts the data into the information. This is possible only if the user acknowledges the issue thoroughly and has extensive knowledge in the data generation. The user should know how to interpret the data that is being generated or developed and should know how best in order to use them.

1.9. ADVANTAGES OF REMOTE SENSING TECHNOLOGY

- Large area coverage: Remote sensing permits the coverage of very large regions which permits the regional surveys on a variety of themes as well as the identification of the features that are extremely large in size.
- Remote sensing permits the repetitive coverage which comes in handy, at the time of data collection on different and vibrant themes. For example, water, agricultural fields and so on.
- Remote sensing permits for easy gathering of data over a diversity of scales as well as the diversity of resolutions.
- A single image captured with the help of remote sensing can be examined and interpreted for use in several numbers of different applications and purposes. There is no restraint on the degree or amount of information that can be collected with the help of a single remotely sensed image.
- Remotely sensed data can easily be processed and analyzed at a much higher pace, with the help of a computer and the data utilized for several numbers of different purposes.
- Remote sensing is unobstructive in nature, more especially, if the sensor is passively recording the electromagnetic energy that

is reflected from or emitted with the help of the phenomena of interest. In simple words, this means that passive remote sensing does not disturb the object or the area of interest.

- The data which is gathered with the help of remote sensing is analyzed at the laboratory. And that minimizes the work that is required to be done on the field.
- Remote sensing permits for map revision at an extensive scale that ranges from a small to medium scale which makes it a bit cost- effective and faster.
- Color composite can be attained or produced with the help of three separate band images, that makes sure, the details of the area are far much more defined when they are compared with only a single band image or aerial photograph is being reproduced.
- Remote sensing is very easy, in order to trace, detect, or find the situations of floods or forest fire that has spread over a large region. And in addition to this, this makes it much easier, in order to plan a rescue mission fast and at its earliest.
- Remote sensing is relatively cost- effective in nature, as well as remote sensing is a constructive method, which is used in reconstructing a base map in the absence of detailed land survey methods.

1.10. DISADVANTAGES OF REMOTE SENSING

- Remote sensing is a fairly costly method of analysis, more particularly when it comes to measure or analyze the regions that are small in area.
- Remote sensing necessitates a different kind of training, in order to analyze the images. In this way, it is very costly with respect to the long run to use remote sensing technology since the extra training must be accorded to the users of the technology.
- It is costly, in order to analyze the repetitive photographs. If there is any type of requirement is needed, to analyze the several numbers of various factors or aspects of the photography features.
- It is humans who select what sensor wants to be used, in order to gather the data, specify the resolution of the data and correction or adjustment of the sensor, opt the platform that will transport the sensor and finds out when the data will be composed. And due

to this reason, it is much easier to introduce human error in this type of analysis.

- Powerful active remote sensing systems, for example, radars that produce or release their own electromagnetic radiation can be disturbing and impact the phenomenon that is going to be investigated.
- The instruments that are used in remote sensing, every so often, can be un-calibrated and this will be led to un-calibrated remote sensing data.
- Every so often, various phenomena that were being analyzed can give the same impression through the course of its measurement which may take into the direction towards a classification error.
- The image being analyzed may sometimes be interfered with the help of other phenomena that are not being measured and this should also be accounted for through the course of the analysis.
- Every so often, the remote sensing technology is oversold to the point where it feels like it is a panacea that will offer all the solution and information in order to conduct physical, biological as well as scientific research.
- The information, which is provided with the help of the application of the remote sensing data may not be complete or it may be temporary in nature.
- Every so often, large scale engineering maps cannot be prepared with the help of the data which is being received from the satellite, as an outcome, it makes remote sensing data collection incomplete.

1.11. GEOGRAPHICAL INFORMATION SYSTEMS (GIS)

Geography is the study of the feature of the Earth and patterns of their fluctuations in the spatial location and time. Several numbers of questions about the agricultural production are geographic in nature because the production relies on the environment and prevailing the socio-economic setting, and hence, both of which fluctuate spatially and in time.

For instance, the queries are related to the management of the natural resources, precision agriculture, agroecological categorization for the planning of the use of land, regional patterns and trends with respect to the technology adaptation, agricultural productivity as well as income, non-profit source pollution from the lands of agriculture, and so on.

Providing the answers to these sorts of question necessitate the access to the large volume of multidimensional geographical (which is spatial) information of weather, soils, topography, water resources, socio- economic status, and so on.

In addition to this, answers to even apparently simple queries necessitate that, the data or information from various numbers of numerous sources to be combined in a consistent form.

GIS, which is the short form for Geographical Information Systems, allows the representation as well as a combination of these kinds of spatial data or information. In order to present the geographical information, conventional methods are being used in two dimensions is in the form of maps. In simple words, maps are the graphical representation of the surface of the Earth on a plain paper.

> Maps give the shape, the way an individual visualize, assess and analyze the geographical (or spatial) information. A map is consisting of points, lines, and area elements which are positioned with reference to a common coordinate system (it is generally the latitude and longitude). Maps are drawn in order to specify the scales and projections.

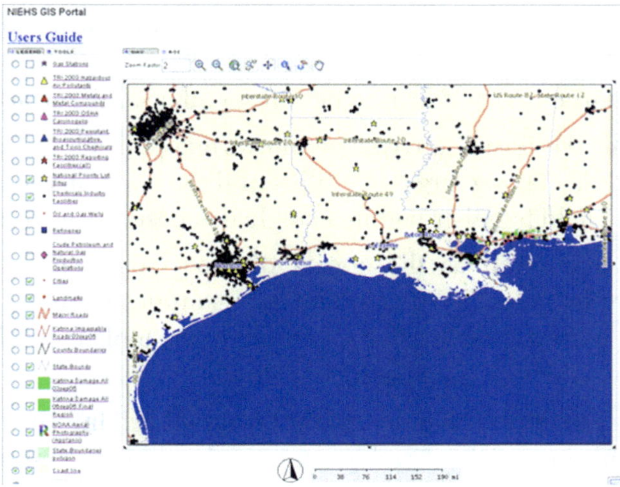

Figure 1.6. GIS through mapping.

Source: Image by flickr

The scale of maps can fluctuate and they rely on the cause for which the maps are developed. In addition to this, projection is a mathematical transformation that is being implemented in order to represent the actual three- dimensional spherical surface of the Earth in two dimensions on a plane sheet of paper.

The legends of the map connect the non-spatial attributes (such as name, symbols, colors, thematic data) to the spatial data. The map itself works as a storage and presents data to the user. For example, analog maps (on paper) are cumbersome in order to develop and implement. More particularly, when there are large numbers of them to be used for the analysis.

Geographical Information Systems are based on the computer application. Geographical Information Systems makes it much easier in order to, both, development of maps as well as the implementation of the maps with respect to several numbers of different complicated analysis.

Geographical Information Systems permits working with the geographic data in a digital format in order to provide assistance in decision making with respect to resource management. Geographical Information Systems is a generic term that depicts the application of the computers in order to develop and display the digital maps.

The attributed data which explains the several numbers of the aspects well as the characteristics in maps, may relate to the physical, chemical, biological, environmental, social, economic or other earth surface properties.

Geographical Information Systems allows mapping, modeling, querying, analyzing and displaying large quantities of these sorts of varieties of data, all of them are held together within a single database.

The power as well as the appeal of the Geographical Information Systems, from its potential to combine the quantities of information about the environment and the extensive repertoire of the tools it provides in order to discover the wide- ranging data.

The history of the development of the Geographical Information Systems parallels the history of the developments in the digital computers and the database management systems on one hand and those in cartography and the automation of the map production on the other.

The development of Geographical Information Systems has also depended on the innovations that are being made under the various sorts of other disciplines as well, such as civil engineering, geography, statistics, photogrammetry, remote sensing, and so on.

A Geographical Information System produces maps and reads maps. The main advantage of the Geographical Information System is that it allows the identification of the spatial relationships between the particular kinds of various features of the map.

Geographical Information System can develop maps in several numbers of different scales, projections and colors. But, in addition to this, is Geographical Information System is just not a tool for map making.

Geographical Information System is majorly a toll for analysis that offers noble ways of looking at, linking and evaluating the data with the help of projecting tabular data into the maps and combining the data which is collected or gathered from several numbers of various sources.

This is being performed with the help of permitting the development of a set of maps in which each is equipped with a different theme (such as soil, rainfall, temperature, relief, water sources, etc.)

From its early beginnings, Geographical Information System has been an integrating technology both from the point of view of its development as well as its application or implementation.

And the reason behind this is that, once the geographic information of any type is modified into the digital form in a Geographical Information

System, it becomes very easy in order to copy, edit, evaluate, manipulate and transmit it.

Geographical Information System permits an important association to be made among apparently unrelated activities that are based on a common geographic location.

This has taken into the direction of fundamental modifications, in such a way that, resource management decisions are made in several numbers of various settings, such as forest management, marketing management, utility management, transportation, as well as in agricultural, environmental and regional planning and management.

1.11.1. Definitions of Geographical Information System

Basically, a Geographical Information System is an information system which works on the basis of computer application, just similar to the other database, but with vital modifications, such as all information in the Geographical Information System must be connected to a geographic (which is spatial) reference (which are latitude, longitude, or other spatial coordinates).

There are several numbers of various definitions of the Geographical Information System. And the reason behind this is that several numbers of users stress various factors or characteristics of its application or use. For instance:

- ESRI defined Geographical Information System as a planned or systematic assortment of computer hardware, software, geographic data and personnel designed in order to effectively capture, store, update, manipulate, analyze as well as display the geographically referenced information.
- ESRI also provided a simpler definition of Geographical Information System as a computer system which is having the potential holding and using that data in order to describe several places on the surface of the Earth.
- Ducker defined Geographical Information System as a special case of information systems where the database is consisting of the information or observations on spatially distributed aspects, properties, activities or events, which are definable in space as points, lines or areas.

A Geographical Information System uses the data about these points, lines or areas in order to retrieve the data for ad hoc queries as well as analyses. The United States Geological Survey (USGS) defined provided a Geographical Information System as a computer hardware and software system that is designed in order to gather, organize, evaluate as well as display the geographically (spatially) referenced data. This definition is fairly comprehensive and is suitable for agricultural applications of the Geographical Information System.

1.11.2. History of GIS Development

The idea of representing the several numbers of layers of data on a sequence of base maps, the connecting things geographically, has been around much older, when they are compared with the invention of computer systems. Hundreds and thousands of years ago, the early man used to draw pictures on the walls of caves, of the animals they hunted down.

These drawings of the animals are track lines and tallies thought to show the migration routes. On the other hand, when they are compared with respect to modern technologies, these early records mimic the two-element structure of modern geographic information systems, which is an image that is linked to the attribute information.

Possibly the earliest use of the geographic method, in the year of 184, John Snow represented a cholera outbreak in London with the help of points in order to represent the locations of several numbers of individual cases. According to the study of John Snow, which depicts that the distribution of cholera takes to the source of the disease, which is a contaminated water pump within the heart of the cholera outbreak.

On the other hand, the basic component of topology and theme existed previously in cartography, the John Snow map was distinct, with the help of methods of cartography, not only to represent but also, in order to analyze, clusters of the phenomenon which are dependent on the geography, for the first time ever.

The time period of the early 20th century saw the development of "photo lithography" where the maps were divided into several numbers of layers. Computer hardware development encouraged with the help of nuclear weapon research takes into the direction to general -purpose computer "mapping" applications by the time period of the early 1960s.

In the year of 1962, the first true operational Geographic Information System of the world was created with the help of the Department of Forestry and Rural Development in Ottawa, Canada by Dr. Roger Tomlinson.

Back then, it was known by the name of CGIS, which is the short form of the Canada Geographic Information System. And, in addition to this, the Canada Geographic Information System was used in order to store, analyze, and manipulate the data which is gathered for the Canada Land Inventory (CLI).

It is an initiative in order to find out the capability of the land with respect to rural Canada with the help of mapping information about soils, agriculture, recreation, wildlife, forestry, and land use at a scale of 1:50,000.

Canada Geographic Information System was the first "system" in all across the globe. Canada Geographic Information System was an enhancement over the "mapping" applications because it provides the capabilities for overlay, measurement, and digitizing or scanning.

1.12. COMPONENTS OF GIS

Geographical Information System allows the user in order to input, organize, manipulate, evaluate, as well as display the geographically referenced data with the help of an application that is based on a computer.

In order to perform several numbers of various operation or functions with the help of Geographical Information System, the components of the Geographical Information System such as software, hardware, data, people and methods are very important.

1.12.1. Software

Geographical Information System software offers the functions as well as the tools which are required in order to store, analyze, and display geographic information. The main components of the software are mentioned below:

1. A database management system (DBMS)

- Tools, in order to input and manipulate geographic (spatial) information
- Tools that aid with respect to the geographic query, analysis, as well as visualization
- A graphical user interface (GUI) in order to facilitate the access of tools.

Geographical Information System software is either commercial software or software developed on open source domains, which are available for free. Never the less, the commercial software is copyright protected, and this software can be expensive in nature and they are available with respect to the number of licensees.

In the present interval of time, the different commercial Geographical Information System software that are available emphasize on Arc / Info, Intergraph, MapInfo, Gram++ etc. among them, Arc/Info is the most popular software package. In addition to this, the open- source software includes AMS / MARS and so on.

1.12.2. Hardware

Hardware is the computer on which a Geographical Information System operates. In the present interval of time, Geographical Information System operates on an extensive range of hardware types, it ranges from centralized computer servers to the desktop computers used in stand-alone or networked configurations.

1.12.3. Data

Talking about the data, it is the most significant component of a Geographical Information System. Geographic data or Spatial data as well as other related tabular data, can be gathered in-house or bought from a commercial data provider. Spatial data can be in the form of a map/remotely-sensed data, for example, satellite imagery and aerial photography.

These forms of the data must be adequately georeferenced (which is longitude or latitude). Tabular data can be in the form attributed to data that is in some way linked to the spatial data or geographic data. Most of the Geographical Information System software are equipped with inbuilt Database Management Systems (DBMS) in order to develop and maintain a database in order to order to aid in organizing as well as managing the data.

1.12.4. Users

The technology of the Geographical Information System is of limited value without the users who control the system and in order to create plans in order to implement them.

The users of the Geographical Information System ranges from technical professional, who design and maintain the overall functioning of the system to the users who use the Geographical Information System in order to help them in doing their work on a daily basis.

These sorts of users are extensively interested in the outcomes of the analysis and can have no interest or knowledge of the methods which are used in performing the analysis. The interface of the Geographical Information System is quite user-friendly which permits the non-technical users to have easy access to the Geographical Information System analytics potential without needing to know the detailed software commands.

A simple User Interface (UI) can be consisting of several menus and pull-down graphics windows, in such a way that, the user can perform the needed analysis with a few key presses without needing to learn specific commands in detail.

1.13. FUNCTIONS OF GIS

The general function of the Geographical Information System software performs six important tasks, like, input, manipulation, management, query and analysis, along with the visualization.

1.13.1. Input

The important input data with respect to any kind of Geographical Information System is the digital maps, images, spatial data as well as the tabular data. Typically, the tabular data is typed on the system of the computer with the help of relational database management system software. Before the geographic data can be used in a Geographical Information System, it must convert into a suitable digital format.

The Database Management System can develop several numbers of different objects, for example, index generation on the data items, in order to increase the rate of the information that is being retrieved by a query. Maps can be converted into digital form or digital representation with the help of a vector format in which the actual map points, lines, and polygons are stored as coordinates.

Data can also be input in the form of the raster, in which data elements are stored as cells in a grid structure (the technical details are presented in the following section). The process of transforming the data from paper maps to the files, stored in a computer, is known as digitizing.

The modern Geographical Information System technology has the potential in order to automate this process entirely for the large projects, on the other hand, smaller tasks can necessitate some sort of manual digitizing. The process of digitizing is labor -intensive and more time consuming, in this way, it is much better to use the data which is already present in the computer system.

In the present interval of time, several numbers of geographic data already present in the formats which are compatible with Geographical Information System. These sorts of data can be acquired from the data suppliers and loaded directly into the Geographical Information System.

1.13.2. Manipulation

Geographical Information Systems can store, allocate, as well as update spatial data which is linked with the text data. The spatial data must be referenced with respect to a geographic coordinates system (which includes longitude or latitude). The tabular data is linked with the spatial data, and it can be manipulated using the database management software.

It is likely that the data types are needed for a particular Geographical Information System project will require to be converted or manipulated in some way in order to make them compatible with respect to the systems. For instance, geographic information is available at various scales (for example, scale of 1:100,000; 1:10,000; and 1:50,000).

Before these can be overlaid and combined, they must be converted into the same scale. This could be a temporary conversion in order to display or a permanent one required for the analysis.

In addition to this, there are several numbers of other types of data manipulation that are performed in the Geographical Information System on a regular basis. This is consisting of projection changes, data aggregation, and generalization and weeding out unnecessary data.

1.13.3. Management

With respect to the small Geographical Information System projects, it may be adequate in order to store the geographic or spatial information in the form of computer files.

Never the less, when the volume of the data becomes large and the numbers of users of the data become very less or very few, it has been advised to use a database management system in order to facilitate the storage, organizing, as well as the management of the data.

A database management system is a data management software package which aids in order to control the combined collection of the database objects, for example, tables, indexes, query, along with the other procedures in a database.

There are several numbers of various models of the database management system, but with respect to the implementation of the Geographical Information System, the relational model database management systems will be highly helpful.

With respect to the relational model, the data is being stored conceptually in the form of a collection of tables. In which, each and every table will have the data attributes which is linked to a common entity. Common fields in several numbers of various tables that are being used in order to connect them together with relations. Because of its simple architecture, the relational database management software has been used so extensively.

These the flexible in nature and they have been very extensively used in applications both within and without the help of Geographical Information System.

1.13.4. Query

The stored information either in the form of spatial data or linked tabular data can be retrieved using SQL, a short form of Structured Query Language. According to the type of interface of the user, the data can be questions with the help of Structured Query Language or a menu- driven system can be used in order to retrieve the map data.

1.13.5. Analysis

Geographical Information systems really come into their own they are being applied or implemented in order to analyze the geographic data or spatial data. The processes of the geographic analysis every so often are known as spatial analysis or geo-processing uses the geographic properties of features in order to find out the patterns and trends, and to undertake "what if" cases.

In the present interval of time, the modern Geographical Information System has several numbers of powerful analytical tools in order to evaluate or analyze the available data.

1. Overlay Analysis

The combination of several numbers of various data layers is consisting of a process which is known as overlay. In simpler words, this could be a visual operation, but analytical operations necessitate one data layer or more than one data layer, which are required to be joined manually. This overlay, or spatial join, can combine the data on soils, slope, and vegetations, or land ownership.

For instance, data layers with respect to the soil and land use can be integrated, and as an outcome, in a new map which is consisting of both soil as well as the information on land use. This will be very insightful, in order to acknowledge the several numbers of various behavior of the situation with respect to various parameters.

2. Proximity Analysis

Geographical Information System can also help in buffer generation that is consisting of the development of the new polygons from points, lines, and polygon features that are stored in the database.

For instance, in order to find out the answers to the queries such as: How much area covered within 1 km of water canal? What is the area covered under several numbers of various types of crops? And, for watershed projects, where is the boundary or delineation of a watershed, slope, water channels, various kinds of water harvesting structures are needed, and so on.

1.13.6. Visualization

Geographical Information System can offer hardcopy maps, statistical summaries, modeling solutions, along with the graphical representation of maps for both spatial data as well as tabular data. With respect to the several numbers of geographic operations. In the end, the outcome is best visualized in the form of a map or a graph.

Maps are very effective in order to store and in communicating geographic information. Geographical Information System offers new as well as interesting tools in order to extend the art of visualization of output information to the users.

1.14. TECHNOLOGY USED IN GEOGRAPHIC INFORMATION SYSTEM (GIS)

1.14.1. Data Creation

In the present interval of time, the modern Geographic Information System technologies use digital information, in which several kinds of digitized data creation methods are being used. Out of which, the most commonly used method of data creation is the digitization.

In the context of digitization, a hard copy of the map or a survey plan is converted into a digital medium with the help of the application of a design program which works on the computer, and in addition to this, the program is equipped with geo-referencing capabilities.

With the help of extensive availability of the rectified imagery (both from the satellite and aerial sources), heads-up digitizing is becoming the main field with the help of which geographic data is abstracted. Heads-up digitizing is consisting of the tracing of geographic data directly on the top of the aerial imagery, in spite of the help of the conventional approach of tracing the geographic form on a separate digitizing tablet.

1.14.2. Relating Information from Different Sources

If any individual could relate the information regarding the rainfall of a state with respect to the aerial photographs of any nation, he or she might be able to tell that, which wetlands dry up at certain times of that particular year. A Geographic Information System, that can use information from that is collected from several numbers of various resources in several numbers of different forms, can help in order to perform the analysis.

The main requirement for the source data consists of knowing the locations for the variables. Locations may be marked up with the help of x, y, and z coordinates of longitude, latitude, as well as elevation, along with that, it can be done with the help of other geocode systems such as postal codes.

Any kind of variable that can be situated spatially can be stored into the Geographic Information System. Several numbers of various kinds of data in the form of a map can be fed into the Geographic Information System.

A Geographic Information System can also transform digital information, which is already present in the system, which may not yet be in the form of a map, into forms it can identify and use.

For instance, digital satellite images that are generated with the help of remote sensing can be analyzed in order to produce a map, such as layer of the digital information about the vegetative covers. In a similar manner, census or hydrologic tabular data can be transformed into the form of maps, which functions as layers of the thematic information with respect to the Geographic Information System.

1.14.3. Data Representation

Geographic Information System data depicts real- world objects, for example, land use, roads, increment in the digital data. Real -world objects can be separated into two concepts or ideas. Out of which, first is the discrete objects (or a house), and second is continuous fields (which is the amount of the rainfall or increment in the rainfall).

There are two extensive methodologies or tactics that are being used in order to store the data in a Geographic Information System for both of the concepts or ideas. And these two extensive methods are Raster and Vector.

1. Raster

Talking about the raster data type, in this context any sort of digital image. Any individual who is used to with digital photography will identify the pixel as the smallest individual unit of an image. A grouping of these pixels will develop an image, which is unique or different from the commonly used scalable vector graphics, which are the basis of the vector model.

On the other hand, a digital image is related with the output as the representation of the reality, in a photograph or art that is transported or relocated to the computer system, the raster type will show an abstraction of the reality.

Aerial photos are one commonly used form of the raster data, which is having a single function or purpose that is to show a detailed image on a map or for the purpose of digitization.

In addition to this, other raster data sets will be consisting of the information, in the context of the elevation, a DEM (which is the short form of Digital Elevation Model), or reflectance of a specific kind of wavelength of light.

Talking about the Digital Elevation Model, Digital elevation model, map, and vector data, the raster data type is consisting of rows and columns of cells, in which each cell is used to store a single value.

Raster data can be of different forms, it could be an image (which is known as raster images) with each pixel is consisting of a color value. Additional values recorded for each and every cell may be discrete value, for example, land use, or a continuous value, like temperature or a null value if there is no data is available.

On the other hand, a raster cell stores a single value, it can be extended with the help of the application of raster bands in order to represent RGB (short form for red, blue, green) colors, color maps (which is a process of mapping between a thematic code and the value of RGB), or an extended attribute table with one row for each distinct value of the cell.

The resolution of the raster data set is its width of the cell in the ground units. Raster data is stored in the several numbers of formats, which is ranging from a standard of TIF which is based on the file, JPEG format to binary large object (BLOB) data stored directly in a relational database management system (RDBMS), which is almost similar to the other feature classes that are based on the vector.

With respect to the database storage, when the database storage is indexed in a proper manner, it generally permits for quicker retrieval of the raster data but can necessitate the storage of hundreds and thousands of the significantly sized records.

2. Vector

A simple vector map, with the help of each of the vector elements, such as points for wells, lines for rivers, and a polygon for a lake. With respect to the Geographic Information System, geographical aspects or properties are every so often expressed as vectors, with the help of considering these sorts of characteristics as geometrical shapes.

With respect to the common or popular ESRI Arc series of programs, these are clearly known as the shape files. Several numbers of various geographical characteristic are best expressed with the help of several numbers of various forms of geometry.

Points

Zero-dimensional points are used for geographical features that can best be expressed with the help of a single grid reference. In simpler words, it is just a simple location. For instance, the locations of wells, the peak elevations, features of interest, or trailheads. Points carry the least amount of information on these forms of files.

Lines or polylines

The use of one-dimensional lines or the polylines, in order to show or represent the linear features. For instance, rivers, roads, railroads, trails, and topographic lines.

Polygons

Two- dimensional polygons are used for geographical features that cover a specific kind of region of the surface of the Earth. These sorts of features or characteristics can be consisting of lakes, boundaries of the city, boundaries of the park, land uses, or buildings. Polygons depict the most amount of information on the file types.

Each and every one of these geometries is associated with a row in a database, that row explains their attributes. For instance, a database that explains that lakes maybe consisting of the depth of the lake, quality of the water, or level of the pollution. This information can be applied in order to develop a map to explain the specific attribute of that particular dataset.

For instance, lakes could be of different colors, and the color of the lake depends on the level of pollution. Several numbers of various geometries can also be compared. For instance, the Geographic Information System could be made into use in order to find out all wells (or all point geometry) that are situated within the radius of one mile (or near about 1.6 Km) of a lake (which is represented with the help of polygon geometry) that has a high level of pollution.

Vector features can be made to respect the spatial integrity with the help of the application of topology rules. For example, 'polygons must not overlap'. The data of the vector can also be used in order to represent the phenomena which are fluctuating on a regular basis.

Contour lines as well as TIN (short form for Triangulated Irregular Networks) are being used in order to represent the elevation or other values that are altering on a daily basis or frequently. Triangulated Irregular Networks stores the values at point locations, which are linked with the help of lines in order to form an irregular mesh of triangles. The face of the triangles shows the terrain surface.

1.15. GIS AND SOCIETY

Along with the popularization of the Geographic Information System, with respect to the decision making, students have begun to study the social consequences of the Geographic Information System. It has been debated that the production, allocation, application, as well as the representation of the geographic information, are largely connected with the social context.

For instance, several numbers of scholars are worried or anxious about the fact that Geographic Information System may not be misused in order to harm society. In addition to this, other related topics are consisting of the discussion with respect to the copyright, privacy, as well as censorship. A more optimistic social approach with respect to the adaptation of the Geographic Information System is to use it in the form of a tool for public participation.

1.16. CONCLUSION

The technology of remote sensing has enhanced from the balloon photography to aerial photography to multi-spectral satellite imaging. Radiation interaction characteristics of the earth and atmosphere in several numbers of various regions of the electromagnetic spectrum are very useful in order to find out and categorize the features of the Earth as well as the atmospheric features.

A Geographic Information System, is a tool that is based on computer application, with respect to the geographical analysis of the information. It is not simply a digitized mas, nor does it store the maps.

The main purpose of the Geographic Information System is to hold a database of spatial data and attribute or descriptive information with respect to the features on a map which can be used in order to develop the required maps.

The crucial concept of the Geographic Information System is the separation of the spatial or geographic reference information and the attributes or descriptive information of the features of the map for the data entry and the database development, and their association through the course of the analysis.

The application of the technology of remote sensing has been expanded at a much higher pace, for the development of all sectors that also includes the field of agriculture.

The techniques or the methodologies of remote sensing will continue to be very significant and a valuable factor, with respect to the enhancement of the existing system of attaining the agricultural data.

The remote sensing provides several numbers of various platforms for the agricultural survey. Satellite imagery has the distinct potential in order to provide the actual synoptic vies of a large area at a particular point of time, which is not possible for the traditional survey methods and also the process of the data acquisitions are very fast with the help of Geographic Information System when they are compared with the traditional or old-fashioned tactics and techniques.

The significance of the application of the remote sensing with respect to the agricultural sector which is consisting of the pattern of land use, production of crops, crop yield determination, along with the monitoring of the crop.

REFERENCES

1. Aggarwal, S., n.d. PRINCIPLES OF REMOTE SENSING. [eBook] Available at: <http://www.wamis.org/agm/pubs/agm8/Paper-2.pdf> [Accessed 16 April 2020].
2. Briney, A., 2019. An Overview of Remote Sensing. [online] ThoughtCo. Available at: <https://www.thoughtco.com/an-overview-of-remote-sensing-1434624> [Accessed 16 April 2020].
3. Escobar, D., Hunter, P., Bishop, P. and Zerger, D., n.d. Introduction To GIS. [eBook] Available at: <http://www.geogra.uah.es/patxi/gisweb/GISModule/GISTheory.pdf> [Accessed 16 April 2020].
4. Grind GIS-GIS and Remote Sensing Blogs, Articles, Tutorials. 2017. Advantages and Disadvantages of Remote Sensing. [online] Available at: <https://grindgis.com/remote-sensing/advantages-and-disadvantages-of-remote-sensing> [Accessed 16 April 2020].
5. n.d. Basic Concepts of Remote Sensing for Retrieval of Land Surface Parameters. [eBook] Available at: <https://shodhganga.inflibnet.ac.in/bitstream/10603/48605/12/12_chapter%202.pdf> [Accessed 16 April 2020].
6. n.d. Geographical Information Systems (GIS). [eBook] Available at: <https://www.manage.gov.in/studymaterial/GIS.pdf> [Accessed 16 April 2020].
7. n.d. Geographical Information Systems: Concepts. [eBook] Available at: <https://naarm.org.in/VirtualLearning/vlc/downloads/GIS%20concepts.pdf> [Accessed 16 April 2020].
8. n.d. MODULE – 1 LECTURE NOTES – 1 BASIC CONCEPTS OF REMOTE SENSING. [eBook] Available at: <https://nptel.ac.in/content/storage2/courses/105108077/module1/lecture1.pdf> [Accessed 16 April 2020].
9. Joseph, 2004
10. Colwell, 1956
11. Gaspard Tournachon, 1859

Chapter 2

Application of Remote Sensing for Environmental Monitoring

CONTENTS

This chapter is about the application of remote sensing. Which tells the use of remote sensing methods in different fields? As remote sensing information can be collected through ground- based manuall means or it can be space -based like using Landsat and many other technologies for monitoring and assessments.

This chapter gives a description of the use of the remote sensing method and its importance. As this chapter will explain all the applications of remote sensing like urban mapping, hydrological use, soil, water and drought detection, information regarding these topics is being provided in a descriptive way.

2.1. INTRODUCTION

Remote sensing (RS) is defined as an art and science of acquiring information like spectral, temporal and spatial, about a material object, phenomenon or area that is without coming into physical contact with the objects, phenomenon or area which is under investigation.

Remote sensing often acquires information about the features of Earth's surface features without coming physically in contact with them. Sensors like scanners, radiometer, cameras and radar -like products are mounted on the satellite or airplanes that record electromagnetic energy. This energy gets reflected or emitted from the surface of the Earth.

The amount of energy that gets reflected is considered to reflect at a specific wavelength. This Wavelength basically depends upon the properties of the medium and optical characteristics of the surface. Thus, the differences that take place between the features that have different optical properties. These properties are being made by analyzing remote sensing data.

This sensing data records the reflectance; this reflectance is being recorded in various wavelength regions of the electromagnetic spectrum. Plots of wavelength versus reflectance or 'spatial signatures' are being used for identifying the areas that are similar to the spectral response on the surface of the Earth.

Image processing software is being used to distinguish between the different spectral signatures that are easily available in the market areas. Then are used to map the extent of different Earth's surface features. From the last few years, it has been noticed that remotely sensed data have been used increasingly as a tool for a wide variety of applications.

This variety of application is ranging from disaster management, to monitoring the natural resources, to health management and disease control.

The recent advances in technology and the availability of remote sensing data are at an affordable price. Through this data, there is a possibility to have frequent monitoring of the possible environmental changes.

It is being observed that the higher spectral, spatial and temporal resolutions of the satellite data are considered to be best suitable for the continuous mapping and monitoring of the environmental parameter, which can be achieved at both the levels of regional and local as well.

The data which is being achieved by the method of remote sensing can be used for changing the detection studies, by simply comparing the current situation with the past one. RS data is not only good for past and present data, but this method is also helpful for predicting the risk in the areas. They are also used for selecting the remedial measures by simply providing the baseline for the future forecast.

There are several other techniques like Satellite remote and Modern airborne sensing techniques. These techniques are offering attractive opportunities for the coastal monitoring system.

2.2. REMOTE SENSING PRINCIPLES

Remote sensing is considered as a new tool in landscape ecological research which is helpful in enhancing the recognition of patterns. This technique is also used for the analysis of the state and recognition of the landscape properties, as well as the trajectories of land escapes.

In a combination with the Geographical Information System (GIS), modern spatial analysis methods and biophysical first principles, remote sensing allow a powerful retrieval and interpretation of the pattern and with that, it also processes the landscapes that are at a range of temporal and spatial scales.

Remote sensing is basically working on the following principles:

- Images of the Earth's surface and of the atmosphere are being sampled from a large or remote distance, which is usually covered from the aircraft or from the satellites.
- It is being recording as a reflected electromagnetic radiation of the Earth's objects and surfaces. This recording is being done either actively by recording the reflectance and emission which is basically from the source that has been projected, or it can be either passively by recording the ground reflected light of solar illumination by using various ranges like visible, microwave or thermal range.

The remote sensing platform is most widely used in landscape ecology. This ecology is based on a passive system. This system works by scanning the near and visible infrared part of the spectrum. This spectrum can range from 400 nm to 1000 nm.

Figure 2.1. Image of a basic remote sensing method.

Source: Image by Free SVG

Therefore, the importance of detecting and recording the reflected radiation originates from the object on the ground which is either reflects or absorbs the incoming radiation selectively. However, they are seen to reflect the incoming energy which is at a variable intensity along with a range of wavelength.

2.3. INFORMATION REQUIRED IN LAND RESOURCES MANAGEMENT

Land resources basically consist of crops, soil, livestock and forests, etc., as the land components of the Earth mineral resources and hydrological cycle which covers soil moisture, underground water, snow cover and associated runoff from the soil. Most of the land resources are seen to use for the production of fodder, food, fiber, fuel wood and for increasing and improving the productivity of the land.

Some of the basic land resources information that is needed is found to be on the soil characteristics, surface and availability of the groundwater, slope and degree of roughness, present land cover and use of characteristics, biological condition. Biological conditions include insect and disease infestations of forest land, grassland and crops, urban development, etc.

There are few important issues that specifically need immediate attention and, in these issues, remote sensing is believed to play a significant role:

- Evaluation of present land -use practices and projections for the future.
- Identification of appropriate corrective and conservative measures required for bringing about the desired production and minimizing the environmental damages.
- Assessment of land resources which are physically useable and economically relevant
- Analysis of constraints related to resource development - physical, economic or social.
- Inventory and mapping of resources.
- Identification of strategies that offer sustained production and other benefits.
- Evaluation of changes in the structure and function of land systems.

2.4. APPLICATION OF REMOTE SENSING

2.4.1. Urban Mapping Applications

It is being noticed that the majority of remote sensing work over the past decades has been focused on a natural environment. It is considered that it is a relatively new method when a remote sensing technique is applied to the urban areas.

The urban remote sensing has been seen rapidly gaining interest in remote sensing communities with the help of the advent of more capable techniques and because of high -resolution imagery. It is being noticed that the remote sensing of urban areas has become a new arena of geospatial technology and it has also application in almost all the socio-economic sectors (Weng and Quattrochi, 2006). This is because of the technological advances and societal needs that are being driven.

Urban landscapes are considered as a complex combination of roads, buildings, sidewalks, parking lots, cemetery, water, soil, garden and many other surfaces. It is being noticed that each of these urban component surfaces is seen to exhibit a unique thermal, radiative, aerodynamic and moisture properties.

They are also seen to relate to their surrounding site environment in order to create a special complexity of an ecological system (Oke, 1982).

In order to understand the dynamic pattern with the processes and their interactions which is present in heterogeneous landscapes like urban areas, it is important that one must be able to quantify the exact spatial pattern of the landscapes and the temporal changes (Wu et al., 2000).

There are some rules that are necessary in order to do so:

- to have a consistent method in order to define theses component surfaces, and
- To detect and map them in repetitive and consistent ways. In this way, a global model of urban morphology may be developed. Monitoring and modeling their changes over time be possible (Ridd, 1995).

The remote sensing technology has been seen to be widely applied in urban land use, change in detection and classification of land cover. However, it seems to be rare when the classification accuracy is achieved to be greater than 80% by simply using the per -pixel classification algorithm which also known as "Hard classification" (Mather, 1999).

In the case of urban areas, the low accuracy of Land used/ Land covered (LU/LC) classification is largely attributed to the mixed pixel problem. In this, there are several types of LU/LC that are considered to contain in one pixel.

The mixed pixel problem is seen to have resulted from the fact that specifies that the scale of observations is called pixel resolution which fails to correspond to the spatial characteristics of the target (Mather 1999).

However, the fuzzy or soft approach of LU/LC classification is being applied. The classification in which each of the pixels is being assigned as a class membership of each of LU/LC type than the single label (Wang 1990). However, Mather suggested that either "soft" or "hard" classification was not considered as an appropriate tool that is being used for the analysis of heterogeneous landscapes.

Ridd (1995) and Mathur (1999), both have maintained the description/ identification/quantification, other than classification, it should be applied to provide an understanding in a better way about the compositions and processes of the heterogeneous landscapes like as in urban areas.

A major conceptual model that is used for remote sensing analysis of the urban landscapes is being proposed by the Ridd (1995), therefore, the soil model (V-I-S) – vegetation – Impervious surface. It is being assumed that the land that covers in the urban environments is considered as a linear combination of three components. These three components are vegetation, soil and impervious surface.

Ridd even believed that this model can also be applied to spatial -temporal for the analyses of the urban morphology, human system and biophysical systems. Meanwhile, urban land uses information that may be more useful for the socioeconomic and planning applications.

Biophysical information is there which can be directly derived from the satellite data which is considered to be more suitable in order to describe and quantify the urban processes and structures (Ridd 1995).

It has been specified that the V-I-S model has been developed for the Salt Lake City, present in Utah. But it is also noticed that this model has been tested in other cities also (Ward et al. 2000, Madhavan et al. 2001, Setiawan et al. 2006).

These studies have been employed by the V-I-S model as a conceptual framework. This is done so because it relates to the urban morphology into the medium resolution satellite imagery. But in this case, it is also being noticed that it is "Hard classification" algorithms that were often applied.

Therefore, it is being noticed that the problem of mixed pixels cannot be solved. With that the analysis of the urban landscapes was still considered to be based on the "Pixels" or "Pixels group".

There is another approach known as Linear Spectral Mixture Analysis (LSMA). This approach can be used to handle the mixed pixel problem, along with that the fuzzy classification. Instead of using any other statistical methods, LSMA is believed to be based on the physically deterministic modeling.

This modeling can unmix the signal that is measured at a given pixel into the components parts which are called endmembers (Adams et al. 1986; Boardman 1993; Boardman et al. 1995). Endmembers are considered as a recognizable surface material that has a specific homogenous spectral property which is present all over the image.

LSMA often assumes that the spectral are being measured by a sensor, which is considered as a linear combination of spectra of all the components that are present within the pixel (Boardman, 1993). Because of its ability to provide the continuum-based biophysical variable and also because of its effectiveness that is used in handling the spectral mixture problem, LSMA is considered to be widely used in the following methods:

- change detection (Rashed et al. 2005; Powell et al. 2007).
- impervious surface estimation and/or urban morphology analysis (Phinn et al. 2002; Wu and Muarry 2003; Rashed et al. 2003; Lu and Weng 2006a, 2006b; Wu et al. 2005),
- estimation of vegetation cover (Asner and Lobell 2000; McGwire et al. 2000; Small 2001; Weng et al. 2004; Lee and Lathrop 2005),
- vegetation or land cover classification (Adams et al. 1995; Cochrane and Souza 1998; Aguiar et al. 1999; Lu and Weng 2004), and

However, these methods are with few exceptions, but these studies have focused on technical specifics. It also focused on the examination of the effectiveness of LSMA.

It is considered that there are only a few studies who have adopted the V-I-S model as their conceptual model. In order to explain the urban land cover patter (Phinn et al. 2002; Wu and Murray 2003; Wu et al. 2005; Lu and Weng 2006a, 2006b; Powell et al. 2007), while with the other implicitly (Rashed et al. 2003, 2005).

2.4.2. Hydrological Applications

The first Earth Resource Technology Satellite, commonly called as ERTS-1 was launced in 1972. Since 1972, scientists have been using the remotely-sensed data that is from the different sensors in order to characterize, analyze, map and model the state of the land surface and processes of the surface.

With the help of this new algorithm, information regarding new hydrological was being extended from the remotely-sensed data. Then this data is being used in environmental and hydrological modeling.

It is believed that these new hydrological parameters and information have been increased the understanding among people about the different hydrological processes by simply helping in quantifying the amount and rate of energy and water fluxes in the present environment.

The basic ability of these sensors is to provide various spatiotemporal scales data. This data has increased the capability of the person to investigate one of the challenges of environmental modeling. With this, it also mismatches between the scales of environmental process and available data.

It is considered that the role of remote sensing for understanding the hydrological processes and fluxes across the different temporal and spatial scales that can be tremendous. This can be possible only when there are appropriate temporal and spatial resolutions for the remotely sensed data is found to be available and that is under a range of bands.

With the help of the availability of a large volume of remotely sensed data, the geographical information systems (GIS) tools are being used in order to manipulate, retrieve and process such data. With that, it also computes the system efficiently, whereas, the application of the remote sensing towards water resources is believed to increase in recent years.

It is noticed that the application of remote sensing which is present in the water resource research and management is considered to lie in one of the three categories, these categories are as follows: Mapping of watersheds and features, direct estimation of hydrological variables, and indirect hydrological parameter estimation.

- Mapping of Watersheds and Hydrologic Features

For different types of sensors that use an airplane or satellite that is being used extensively for providing images, photographs and providing mapping information that can be used for different purposes.

Mapping includes the area of wetland, disaster areas, riverbanks, snowpacks, flood plains, coastal shores, drainage basins, fire damage, and many other areas that are showing the areal extent of a given land. These land features may be distinct from other areas due to the presence of difference in the spectral signature that basically falls under these categories.

Different studies have been using aerial photos, LiDAR and radar data, and satellite images for the purpose of mapping and visualizing the land surface for the purpose of planning, hazard assessment, emergency operations and resource mapping.

- Flood Mapping of Devils Lake

Landsat Thematic Mapper and Enhanced Thematic Mapper Plus (ETM+) images are being aboard by the Landsat-5 and 7 satellites, respectively. These Landsat-5 and 7 satellites have been used from 1991 to 2003, which were used to delineate the flooding map and with this, it also uses to compute the lake surface area.

It was shown that the mosaic of three Landsat scenes containing path/row as 31/27, 32/26, 32/27. These Landsat were used to cover the entire study area. The Lake surface area of the chain of Devil lakes is considered to exclude the Stump Lake that is being mapped by the Landsat in the years 1993 and 2003, which is indicated in figure 2.2.

In the figure it can be seen easily that between the years 1993 and 2003, the surface area of the lake was quite low in the year 1993, whereas in 2003 it shows that it has the highest area. For each respective years, the lake surface area images taken or received from the Landsat were more classified and in this only lake area was computed.

In recent years it is noticed that the image that depicted that the areal increase of the lake is quite greater. In figure 2.2., the white line marked shows the boundary of the lake line in the year 1992. The consistency in the surface areas is also shown in figure 2.2. From 1993 and 2003, which are indicating that there is an increase in the flooded area which is about 117% from the years 1993 and 2003.

There is another figure 2.3. This shows that lake surface area year by year which does not contain Stump Lake. This surface area is derived from the

Landsat images which shows that each of the meters rises of the Lake which inundates about the progressively greater area which in return increases the volume of the flood.

In this figure 2.3., it is shown that there is a 1.3m of increase between the years 1993 to 1995 with an elevation of 435.8m which adds about 39.5km^2 to the lake surface area. In this area there is a 1.3m increase takes place between 1996 to 1997 with an elevation of about 438m which adds near about 49km^2 (Melesse et al. 2006a, 2006b).

Therefore, this indicates that there is the severity of flooding problems that are at the higher lake levels which are being attributed partly due to the flat topography which is contributing to the watershed.

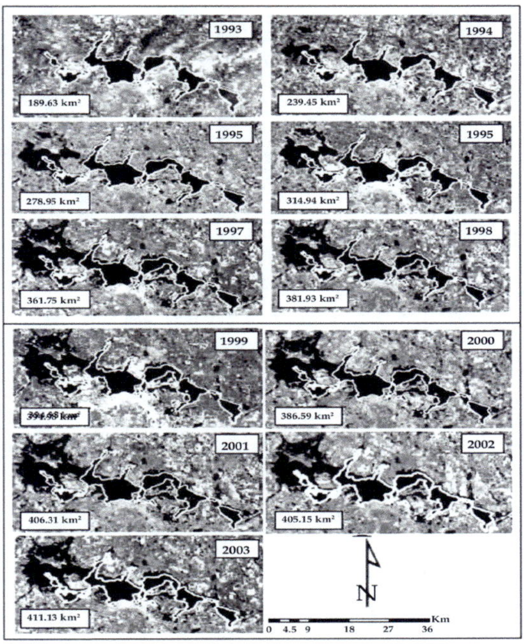

Figure 2.2. The spatial expansion of Devils Lake from Landsat images.

Source: Image by Assefa M. Melesse et al., 2007

- Indirect Estimation of Hydrological Parameters

It is considered that there are a majority of remote sensing contributions that are present in the water resource management which falls under the indirect hydrological parameter category estimation. In this category, it is considered that there is the most common area of the contribution that focuses on the use of classification algorithms in order to generate the classes related to land cover.

By using this unique spectral signature of land surfaces which mainly includes visible, thermal and infrared spectra in order to cover the land classes which help in estimating such data. In order to demonstrate the mapping land cover for the Heart River sub- basin, this is present in the Missouri River Basin in North Dakota. They basically use the surface temperature, unsupervised classification algorithm and Normalized Difference Vegetation Index (NDVI) for demonstrating.

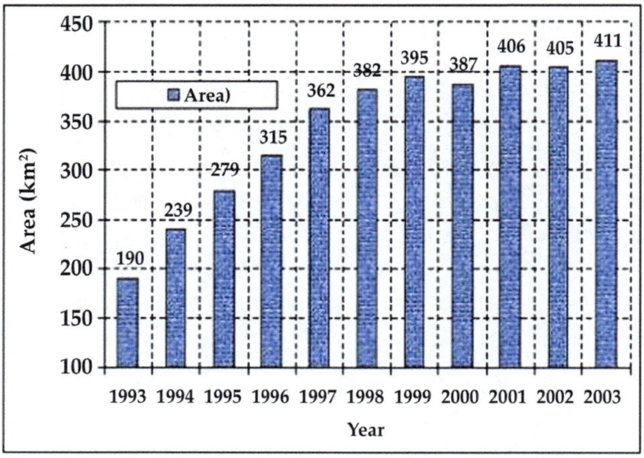

Figure 2.3. Devils Lake surface area progressions from Landsat images.

Source: Image by Assefa M. Melesse et al., 2007

- Land Cover Mapping

The Land cover for the Heart River Sub-basin of the North Dakota was determined on July 2, 2002, through the Landsat ETM+ and the images were obtained by using the ISODATA classification (ERDAS, 1999). There are some unsupervised classifiers that yield about 30 spectral classes.

Scattergram of the Scaled Normalized Difference Vegetation Index (NDVIs) when compared to Scaled Radiant Temperature (Ts)were considered to be used in order to find the instances of strong correlation that basically take place between them and the land cover data of the sub-basin.

The spectral often signatures about all these classes, where it is being used, to determine the mean radiance for each of the band. By using this mean signature values, there should be additional layers and vegetation indices can be derived.

By using the scatter gram, five United states Geological surveys were the Land covered and Land used by using the system level 1land cover classes (Anderson et al., 1976) were basically identified (Melesse, 2004a).

- Fractional Vegetation Cover

In order to understand the changes that took place in the vegetation basically covers the images of different scenes and dates. The scale of NDVIs has been used by many of the researchers (Price, 1987; Che and Price, 1992; and Carlson and Arthur, 2000).

$$NDVI_S = \frac{NDVI - NDVI_{low}}{NDVI_{high} - NDVI_{low}} \quad (1)$$

In the above equation, NDVI $_{low}$ and NDVI $_{high}$ are the values that are used for the bare soil and for the dense vegetation respectively. Carson and Ripley (1997) found that the relation between the scale NDVI and fraction vegetation cover (FVC) is to be

$$FVC \approx (NDVI_S)^2 \quad (2)$$

In the above equation, FVC ranges between 0 and 1. From the images of Landsat in the year 1986 TM and ETM+ of 2000, PVC was estimated for the Econlockhatchee River sub-basin (Econ sub-basin), Florida and is being compared in order to understand the change in the vegetation cover that takes place over the 16 years (Melesse et al. 2003a, 2003b).

- Impervious Surface Cover

The surface will obstruct the natural infiltration of the water and which also enhances the surface runoff which is being classified as an impervious surface. When this gets associated with the urbanization and construction of the pavements, building and roads, impervious surfaces will pay as an important role in the surface runoff and for the transportation of the contaminants.

Therefore, the Remote sensing technique has been used as an effective technique in order to map the impervious surfaces by simply using the spectral characteristics of the surface (Melesse 2004b, Melesse and Wang, 2007).

Ridd in the year 1995 and Owen et al., in the year 1998have shown that the relationship between the Fractional Vegetation Cover (FVC) and Fractional Impervious Surface Area (FIS). This relationship is shown for the developed areas as

$$FIS = 1 - FVC \qquad (3)$$

In the case of surface runoff estimation, impervious surfaces are being classified as being hydraulically connected and also those who are not connected.

The hydraulically connected surfaces like parking lots and roads are found to be connected with the drainage system. In this system, there is runoff that takes place from such surfaces and leads towards the drainage network.

Those surface rooftops are being classified as a hydraulically not connected. Whereas, the rainwater from the rooftops can even fall into the pervious surface area, such as grass can be termed as disconnected.

It is considered that the resulting storm runoff which is being gained by such surfaces can be lower than the hydraulically connected impervious surface areas.

- Micro-Topography and Latent Heat Flux

The topography is considered to play an important role in the distribution and in the flux of water and energy that occurs within the natural landscapes. Surface runoff, infiltration and evaporation processes are often taken place at the ground atmosphere interface.

Quantitative assessments of these processes are often considered to depend on the configuration of the topographic of the landscapes. This is considered as one of the several factors who are controlling the boundary condition.

Wetness index (WI) is considered who provides a description of the spatial distribution of the soil moisture by using the topographic information. WI is calculated as:

$$WI = Ln(A/S) \qquad (4)$$

In the above equation, the witness index is considered to control the flow accumulation, distribution of saturation zones, soil moisture, and depth of water table, thickness of the soil horizon, pH, organic matter, silt, sand content, plant cover distribution and evapotranspiration (Kulagina et al., 1995; Florinsky, 2000).

2.4.3. Soil, Water and Drought Monitoring for Early-Warning Applications

It is believed that with the advent of grid -based, which is remotely sensed about the rainfall data. The application which is being used for the crop water balance model for having crop monitoring and forecast is being yielded, which has gained an increase in the acceptance by various national, international, and local organization that are present around the world.

Soil water is considered as a key state which is variable in the hydrological modeling and which also determines the partitioning of the rainfall into deep percolation and runoff. It also helps in controlling the rate of evapotranspiration (ET).

Although it is considered that the estimation of the actual evapotranspiration (ET_a) is an ultimate goal for many researchers in order to achieve the hydrological and agronomical application. But it is also found to be difficult to quantify and to provide the expensive instrumentation.

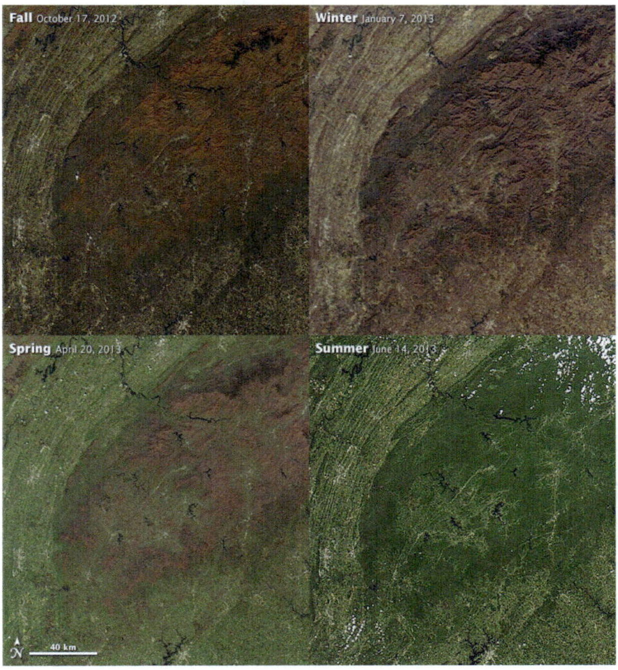

Figure 2.4. Image representing image of a landscape using remote sensing method in different seasons.

Source: Image by NASA Earth Observatory

Therefore, the hydrological modeling is considered as a technique which is being used for estimating ET_a which is based on energy balance principle (Bastiaanssen et al, 1998; Allen et al, 2005; Senay et al, 2007a) or it is water balance based algorithms (Allen et al, 1998; Senay and Verdin, 2003).

In order to monitor large areas by using remotely sensed data, it is believed that the water balance approach helps in providing an operational advantage that is available in terms of data availability.

Meanwhile, the energy balance models are being driven by the thermal data. Whereas, the water balance model, has been achieved by the rainfall.

Naturally, the cloud cover is considered an issue that basically provides a daily estimate of ET_a which depends on the rain -fed agriculture from the model of energy balance. On the other hand, it is noticed that the satellite-derived rainfall data helps in achieving it by various temporal and spatial scales.

This helps in making the various estimation of ET_a by simply using a model of water balance which is a relatively easy task for making various decisions in the field of natural resources and in agriculture.

The most effective and widely used water balance technique that is used for operational use is the FAO water balance algorithm. This algorithm produces the crop Water Requirement Satisfaction Index (WRSI), which can be also called a Crop Specific Drought Index (CSDI).

WRSI is seen to show the relative relationship in the form of ratio/ percent, which takes place between the supply that is attained from the rainfall and soil moisture which is already available, and the demand which is attained when the crops demand to meet their physiological needs.

This method is achieved by using the observed data that is being collected at the beginning of the crop season until the current date data.

If the value of 100 comes as a result then that indicates that all the crops demand has been met. But when the value comes less than 50, then that generally tells that a severe water shortage is there which can lead to complete failure of the crop (Smith, 1992).

When the value comes between 50 and 100, then that will indicate the different degrees of crop stress and reduction in the yield from the shortage of adequate supply. FAO in their studies (Doorenbos and Pruitt, 1977) have shown that WRSI can be also related to crop production by simply using a liner yield reduction function that is specific for a crop.

Meyer et al., in 1993 helped in enhancing the concept of crop water balance modeling. They have used crop stage -specific drought sensitivity coefficient specifically for the corn crop. The Famine Early Warning System Network (FEWS NET) has demonstrated a regional implementation of FAO WRSI in the grid cell modeling environment (Verdin and Klaver, 2002).

Later, Senay and Verdin in 2003 enhanced the geospatial model by presenting the soil water stress factor that is achieved from the process of irrigation engineering, in order to have a better estimation of ET_a as a function of soil water content and the concept of Maximum Allowable Deletion (MAD).

The current version of the USGS/FEWS NET geospatial WRSI model is always found to be operational with daily and 10-day outputs.

The seasonal crop water requirement satisfaction index (WRSI) for a crop is totally based on the supply of water and also on the demand that is raised by the crop which it experiences during a growing season. It is often calculated by the ratio of seasonal actual evapotranspiration (ET_a or supply) to the seasonal crop water requirement (ET_c or demand):

$$WRSI = ET_a * ET_c^{-1} \qquad (5)$$

ET_c in the above equation is calculated from the product of the Penman-Monteith in reference to the evapotranspiration (ET_0) by using the standardized FAO equation which uses short grass a reference crop (Allen et al., 1998) and crop coefficient (K_c).

Crop coefficients, piecewise linear weighting functions are considered to have traditionally been used to adjust for the type and growth stage of the crop:

$$ET_c = Kc * ET_o \qquad (6)$$

USGS/EROS has been generated in order to have operational daily global ET_0. Since 2001, there is a use of this meteorological forcing from the Global Data Assimilation System (GDAS) (Senay et al. 2007b).

ET_a basically represents the actual amount of water that is withdrawn from the soil water reservoir or Bucket. The important difference between the ET_c and ET_a is that ET_a is seen to depend on the soil moisture, which is being calculated on a daily basis in order to provide the daily soil stress correction factor, which is K_s (Senay and Verdin, 2003).

$$ET_a = K_s{*}Kc{*}ET_o \qquad (7)$$

In the above equation, K_c is considered as a crop coefficient, ET_o is known as the reference ET, and K_s is the soil water stress index.

When the soil water content is found above the maximum allowable depletion (MAD) level, which is seen to vary by the vegetation type, then as a result ET_a will help in balancing the ET_c which in return results in no net water stress.

Therefore, when it is observed that there is a fall in the available soil water below the MAD level then the ET_a will be seen to be lower than ET_c.

These changes will be observed in the proportion of the remaining soil water content. The runoff and deep drained out of the root zone are being assumed to occur in the excess of the field capacity.

It is often being noticed that the soil water content is being obtained through an equation of simple mass balance. In this equation the level of soil water is being monitored in a root zone of the soil layer, which is often defined by the Water Holding Capacity (WHC) of the crop root depth and the soil, this equation is represented as:

$$SW_i = SW_{i-1} + PPT_i - ET_{ai} - RF_i - DD_i \qquad (8)$$

In the equation given above, SW is considered as the soil water content, RF is known as runoff, PPT is the precipitation, DD is used to indicate the deep drainage which is being obtained below the root zone, and is the temporal index.

- Model Inputs

Precipitation is considered as a key input data which is used to the water balance model, potential evapotranspiration (PET), and the crop coefficient, and the holding capacity of the soil water. Meanwhile, the key crop coefficient values are being obtained from a FAO publication (Allen, 1998), the other three main datasets are being described below:

- Precipitation

Precipitation is considered as a single most important input of the model. National Oceanic and Atmospheric Administration (NOAA) and the National Aeronautics and Space Administration (NASA) have generated Satellite-Derived Rainfall Estimate (RFE) that is being used to drive the water balance model.

NOAA produces daily rainfall data which is about 10km of spatial resolution by using a blend of Meteosat's Cold Cloud Duration (CCD) and it also produces rain gauge data specifically for Africa (Xie and Arkin, 1997).

NASA is noticed to generate a daily rainfall of about 25 km of resolution from the Tropical Rainfall Mapping Mission (TRMM). TRMM is a satellite system that basically covers the globe between 60 degrees latitude north and south of the equator.

- Potential Evapotranspiration (PET)

USGS/EROS is calculating the daily PET values for the global at 1.0 degrees which is approximately 100 km. This is of spatial resolution from 6 hourly numerical meteorological models with the output fields of the Global Data Assimilation System (GDAS). This is achieved by using the standardized Penman-Monteith equation (Allen et al., 1998).

According to the recent validation of GDAS-PET using PET which is derived from the station parameters in the US which have been demonstrated by the usefulness and reliability of GDAS-based PET for the crop water balance studies (Senay et al. 2007a).

- Soils

According to another important model parameter that is used for the crop water balance model, the soil water is helping in holding water capacity. For the global application, a digital soils map of the world from FAO is used (FAO 1988).

- Operational Model Setup

The USAID Famine Early Warning System Network (FWES NET) is a project at U.S. Geological Survey Center for the Earth Resources Observation and Science (EROS) which has set up operational crop water balance model from the year 2000. This has been done to monitor crop performance in several regions of the world.

These regions are Central America, Afghanistan, Haiti, and Africa. Daily and dekadal (10-day) model outputs are obtained by the crop water balance which is being posted on a website for regions like Central America, Afghanistan, and Africa.

These products along with many other satellite-based monitoring products are being used in order to develop a weekly weather hazard map. These maps are made through a consensus among the partners of FEWS NET. these partners are, USGS, NOAA, NASA, USDA and Chemonics International.

The product which is obtained is being posted regularly as a graphics and in the image data form, which is at the early warning website of USGS.

2.5. OTHER INTEGRATION AND APPLICATIONS WITH RELEVANT DATA SETS

2.5.1. Applications

It is normally seen that the workshop attendees are interested in re-examining the projects in which they are involved. This is done by them so as to see that whether the use of remote sensing data can help to improve the data acquisition and with that it should also facilitate the process of building a powerful information system, especially for their organizations.

It is believed and is being mentioned earlier that more and better amount of data is basically to allow a better understanding of the environmental and human conditions.

This will in turn, lead towards better management of natural resources and with that, it not only focuses on improving the quality of life for better communities, but it also helps to ease the entire decision -making process which is done especially on the government level.

Below there are some key areas that are discussed.

2.5.2. Landscape Assessment

One of the most important applications which include medium resolution sensors like Landsat EMT+, landscape change and is regionally considered as regional landscape assessment. The basic concept of this is that the vegetation features can also be distinguished from other features.

This differentiation is possible only because of their different spectral responses that are in the infrared and visible regions.

It is considered that in the near -infrared bands, vegetation is seen to have a peak of reflectance. This type of reflectance is not commonly seen in the water and urban because of the reduction in the absorption by chlorophyll and other pigments.

Due to the scale at which landscape assessment is basically done in these types of regions, because of this reason it allows the use of Landsat data even from the year 970s and 1980s. In these years, it was made possible even in the absence of TM sensor.

It is believed that the MSS data used for the coarser -resolution will be still useful for the application of a regional scale. Then the priority for the landscape studies will be considered to determine which land cover classes are most important that can be used for identifying on the basis of specific needs.

2.5.3. Air Quality

Air quality application deals with the use of remote sensing for determining the effects of air pollution on vegetation. This is basically useful for urban and sub -urban areas. Changes that occur in the spectral signature of a damaged plant help by giving an indication of nature and about the level at which the air pollution damages.

Such changes can be also detected by simply looking at the changes that occur in the foliage colors and differences that are observed in the texture and structure of the canopy. In the case of small cities or areas, a high-resolution sensor like IKONOS is being required to have a clear picture of the condition of the area.

Assessing the effects of air pollution and with that separating them from the drought or diseases is not an easy task. Disease is considered to affect equally to the trees and they allow them to change the color. By changing the color, it gives a different spectral signature, which can be further misinterpreted as an effect that is caused by air pollution.

In this field, the remote sensing data can be integrated with the field of observation and measures of air pollution.

2.5.4. Water Quality

Pure water often reflects some of the incident radiation which is present in the visible bands of the electromagnetic spectrum. It also absorbs almost all of it which is near and middle infrared bands. It is observed that in the infrared, water often appear as dark and it can easily distinguishable from other features of land.

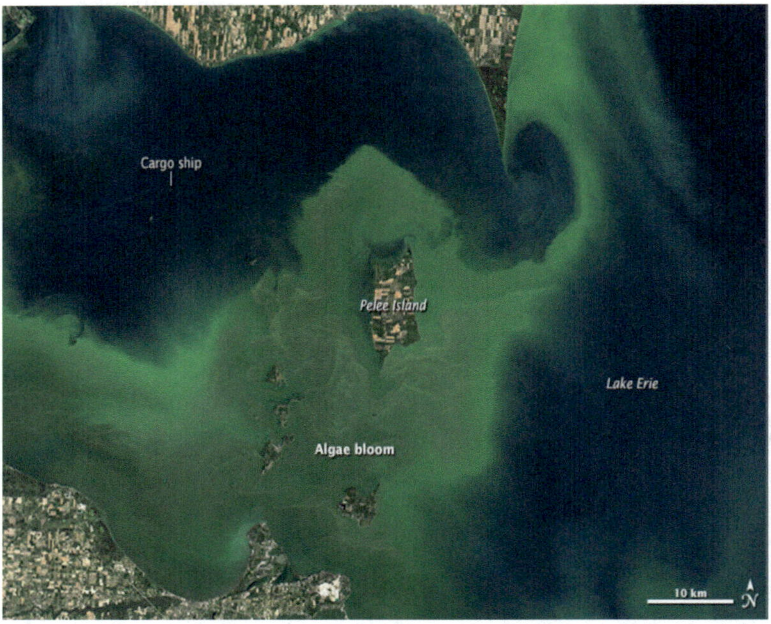

Figure 2.5. An image representing the presence of Algae bloom in the lake.

Source: Image by Landsat - NASA

Spectral response of water can also vary with the presence of suspended sediments. This spectral response will also increase the amount of radiation that gets reflected. According to some of these sediments, it is believed that the suspended solids from the soil erosion can significantly impact the reflectance of spectral and it can be also identified easily in the presence of visible bands.

When there is the use of common satellites like Landsat, the phytoplankton cannot be distinguished easily from the inorganic materials that are present with them. Even there are many more specific sensors that are being required for more complex and sophisticated calibration. Examples of such sensors are SeaWiFS, MODIS, etc.

For example, watershed management and water quality modeling with the traditionally used remote sensing data of soil, land, slope etch can be used as an input for a hydrological model. Such models can be further use in order to estimate the non-point source of water pollution.

There is some non-point source of water pollution that is very common. But this source can be also difficult for detecting the given limitations of traditional *in situ* measurement techniques that can be used for modeling and identifying the sources of pollution.

With the help of integration of in situ measurement with the remotely sensed data which is available in GIS modeling, which can be very useful information on water quality when applicable to many management issues and planning.

2.5.5. Floodplain Mapping and Emergency Management

It is believed that areas like parks, land in urbanized watershed and wetlands are required for emergency management which usually requires a high amount of spatial resolution data that is in a timely fashion.

Reports on remote sensing applications at the state level and local level are in relatively stable areas. In these areas, it can be monitored every one or two years. Although, there are more critical areas that are needed to be monitored more frequently (Lillesand, 2000).

In case of disasters like storms, earthquakes, floods, storms, fires or tornadoes, it is considered that it is very important to have images related to that are both pre- and post-disaster.

Although, it is necessary that pre-disaster is needed to be updated less frequently i.e., every one to five years. It is also important that they are at a comparably high resolution in comparison to post -disaster data.

In the case of disaster management, the satellites with the medium ground resolution are not very useful. For example, Landsat or SPOT. Although, the temporal resolution is relatively higher, which is about 16 to 26 days.

However, the panchromatic and near- infrared aerial photography or IKONOS panchromatic data are considered that they are acquired immediately after a disaster. It also considered as the best way that is used to map the extent of the disaster and to estimate the effects caused by the same. In the presence of cloud, it is believed that it can provide the most useful information.

2.5.6. Health Applications

It is seen that the contribution of environmental factors towards human health and diseases has been recognized and are discussed earlier since the time of Hippocrates. It is believed that before the involvement of scientific explanation that was used for illness, there have been many references to these areas with "bad water" and "bad air".

Therefore, the responsibilities that were for controlling such environmental hazards for the purpose of safeguarding public health has been seen to be increasingly fallen on the local and regional governments. Therefore, there was an official attempt that was made to educate and solicit the inputs and assistance from the local residents.

Contaminated air, soil and water and spread of infectious diseases, when human get in contact with these things then they are considered to be linked with the issues which are taken as environmental health issues. Some of the health concerns seem to be directly related to human activities.

Whereas, other health concerns may be naturally occurring, although they often get indirectly affected by the human -induced changes, example, arsenic in the groundwater or diseases related to vector -borne. In the case of the West Nile Virus and Lyme disease, it is considered that human infection basically depends upon the proximity of arthropod vector, human population and animal hosts.

There are some useful predictors of potential risk, such as land use, land cover and environmental conditions. However, to monitor and control the spread of the West Nile Virus, it was noticed that the local government has tried to map the exact location which includes areas of standing water, in order to find the location of all the dead birds and breeding sites of mosquito.

In order to track the high-risk areas of Lyme disease, the local government focused on the suburban extension where they have found new development turned into heavy vegetation areas. For identifying the areas of high risk, it is considered that the traditional method of field survey will be much costlier and with that, it is also a time-consuming process.

To have more effective results and to focus on selecting more effective areas for surveillance and intervention, local government should utilize this new technique of remote sensing imagery. Unfortunately, it is found that the resolution of the satellite such as Landsat is not sufficient enough that can be used for identifying many of the parameters that are needed to monitor the vector -borne diseases.

Therefore, it can be fine if the use of medium resolution satellite data is in the form of a baseline for many of the major features like large wetlands and vegetated areas. After getting the images, it can be then integrated with either high -resolution satellite or even with the aerial photography data and even it gets supplemented with the data which is received from the ground.

It is considered that remote sensing and GIS can provide regional and local governments with the tools that are useful and with that they also provide information that will help to address these types of health issues that are related to environment.

Even the data which is being collected from the Remote sensing application from the state and local levels, this approach can also be used for determining the routes of exposure from hazardous waste sites and spills.

Even the integrated remote sensing/GIS can also be used for evaluating an alternative site specifically for hazardous facilities like incinerators, which can be used for minimizing the impact on the ecosystem and local population.

Integration that occurs with the relevant data product is basically derived from the satellite imagery that can be easily integrated into GIS by using other data layers that are relevant to specific applications. One of the most useful applications that have emerged during the workshop.

These applications got integrated with the land use and land covered change maps which contain population data derived from the Census.

The main interest behind this is to determine the land cover and land use classes in which there is a change in the population. This change occurs in terms of total population density and population composition.

There is another interesting application that is being used to determine what to the extent and whether to extend the population changes that are occurring in the same areas where land -use changes are occurring.

The integration of remotely sensed data with the data of population can be helpful in relating it to broader topics of urban and sub -urban remote sensing. With this, there is also requirement resolution that is needed by many urban applications.

There are also some spatial resolutions at which sub -urban and urban features and changes can be detected. These changes are more important than the spectral resolution. So, if the data is required frequently than every five or ten years, then ,in that case, IKONOS is the best source that can be used for such data and but this is considered to be quite expensive.

When a calibration of coarser resolution data is achieved from panchromatic Landsat which is with the higher resolution, then this data can solve some of the resolutions related issues, it can potentially provide more coverage and it also reduces the cost.

2.6. MODERN ADVANTAGES OF REMOTE SENSING

Nowadays, remote sensing technology is playing a major role in geography and this is because of the fact that attention is being paid to the latest information, planning as well as management for public and private interests. It is considered to be much useful for natural resource management, environmental degradation, disaster management and sustainable development.

The satellite data of remote sensing is used as basic inputs for the inventory of natural resources as well as development processes such as agriculture, forestry, geology and soil (Chavez et al., 1977).

There are some other important technologies also and these technologies are available to geographers as well. These technologies include maps, SAR, LiDAR, SONAR, aerial photography/photogrammetry/pictometry and GIS. There are some other technologies also which have similarities and differences between them.

2.6.1. Maps

A map is being described by the International Cartographic Union. According to them, 'map can be defined as a conventionalized image which represents the selected features or characteristics of geographical reality.

Maps are designed to be used when the spatial relationships are most important. This is the definition that mentions that in each and every map, there is a scientific accuracy and also, there is a process of selection present (which includes symbolization, abstraction, and generalization).

Nevertheless, in addition to this, the International Cartographic Union says that a map is a thing that shows people the world as they know it. People know that the world is a very complex subject that contains the limits of matter, technology and measurement tools. These are the things which people believe that exists, the people think that are important and the people want and aspire for.

Therefore, the map is subjective because of the fact that the people always decide what to put it and the way in which it should be represented. Contrary to this, a remote sensing image is an objective recording of the electromagnetic which reaches the sensor.

There is one more difference that seems to be important and that is a map is said to be a projection of the earth on paper without any relief displacements, whereas, in a remote sensing image, the map is said to be a projection of relief displacements and geometrical distortions.

2.6.2. Aerial Photography/Photogrammetry/Pictometry

The aerial photography/photogrammetry/pictometry are the systems that collect the data related to the upper surface of the earth and this is done by measuring the electromagnetic radiation from airborne systems. There are some of the major differences which are mentioned as:

The aerial photos are taken by using an analog instrument. The analogs instrument is the film of a photogrammetric camera. After that, it is scanned to be converted into digital media. It has a high resolution (granularity) which is considered to be an advantage of a film.

Figure 2.6. Aerial view of development in "PUKOO"

Source: Image by Wikimedia Commons

On the flipped side, the CCD has the advantage and that is the radiation that reaches the sensor can be measured quantitatively (it is the radiance values, rather than a gray – value scale bar). Therefore, the remote sensing data can further be integrated with physical equations of energy – balance.

2.6.3. SAR

SAR means Synthetic Aperture Radar. SAR tends to provide imagery during the time of night or even in bad weather and also, during the day time.

Images thus produced by using SAR can further be used in order to map the earth's resource and also for environmental monitoring. This requires imaging of the broad-area at high resolutions.

The photographic and some other optical imaging capabilities are complemented by synthetic aperture radar. The reason behind this is the minimum constraints on the time – of – day, atmospheric conditions, and some unique responses given by terrain or cultural targets to radar frequencies.

2.6.4. LiDAR

LiDAR that is Light Detection and Ranging is said to be another technique that tends to offer a number of various advantages in comparison to the other conventional methods which are used for the topographic data collection. Furthermore, this technique provides data which has higher density, higher accuracy, and light independence, less time for data processing and also, it requires some minimum ground control points.

Because of these characteristics, there are some conventional techniques that are complemented by LiDAR in some of the applications whereas there are some other applications also where it completely replaces them. There are a number of different applications where LiDAR data are being used and these applications include geomorphology, forest biomass mapping, glacier studies, and generation of the digital elevation model.

2.6.5. SONAR

SONAR means Sound Navigation Ranging. The SONAR can be considered as remote sensing and this can be done by studying the surfaces of the sea (these features include bathymetry and sea bed features) from a distance. The SONAR is said to be an active type of remote sensing but this technique tends to have sound waves and not electromagnetic radiation, such as Radar, which does not rely on an external source of waves.

The transmission of the waves in both the systems is done by using an interfering medium which can water, air. This then adds up the noise to the data. For the purpose of corrections, the application of these should be done for the purpose of raw data collection.

However, considering the remote sensing, it can be said that the RADAR is considered to be almost weather – independent and atmospheric disturbances tend to majorly affect passive remote sensing.

Both the systems tend to depend on calibration in order to make these required corrections whether it is regarding salinity, temperature, pressure that are measured by the ship while surveying, or even measurements of the atmospheric profile parameters which are done by a meteorological radiosonde.

2.7. FUTURE DIRECTIONS OF REMOTE SENSING

There are certain advancements that are done in the space borne remote sensing if the last 50 years are considered. The advances from Sputnik 1 to Worldview-1, has been phenomenal. As per the present trends, it points towards the increase in a number of various innovations. These innovations include many and from them, seven are mentioned below:

- Advances have been made in the availability of data that are extracted from multiple sensors having a wide array of spatial, spectral as well as radiometric characteristics. These data will be available from a number of different sources.
- Advancements have been done in the availability of data from a constellation such as from Rapid eye which is a very high resolution of around 6.5 meters. These are present in five bands which include a red-edge band.

These have the ability to cover the entire world in 13-day frequency which is considered to be a likely attractive form of data collection. Certainly, this will require innovations in data handling, data storage, and also, in the backup. But for the purpose of applications, a combination is said to be very attractive and that is of very high spatial resolution and frequent coverage.

- There are significant advances that have been made in harmonizing as well as synthesizing data from a number of different sources which tends to facilitate the use of data from these sensors. These sensors have widely differing characteristics and also, different sources. It is also expected that the vendors will market data from multiple sources by the process of harmonizing and by adding value.
- The microsatellites are developed which have the specialty of gathering data for some particular geographic locations and for some specified applications. These microsatellites are likely to become more attractive.

- It is important to have a global wall – to – wall coverage for a number of environmental as well as natural resource applications and here satellites such as Landsat are continuing to play their role which is considered to be very important for the same.
- There are advances that have been made in data synthesis, presentation, and accessibility by using the systems which include Google Earth. These systems will bring in new users and multiply the applications of remote sensing considering the field of environmental sciences and natural resources management.
- Advancements have been done in data availability that is in hyper-spectral and hyper-spatial sensors. These sensors bring in new challenges in the process of data mining, data processing, backup and retrieval.

2.8. CONCLUSION

Earth-observing satellites are the source that provided people with an unprecedented view of the land surface. But these satellites have also exploited relatively less amount that can be used for the measurement of environmental variables that are of particular relevancy with the epidemiology.

According to the recent advances that are occurring in the techniques that are used for recovering the continuous field of air, humidity, vapor pressure and temperature that deficits from the remotely sensed observations. They are considered to have significant potential for the disease vector monitoring and also with the related epidemiology applications.

It is being reported that the development of the techniques that are used to map the environmental variable which is with the relevance to the prediction of intermediate host and relative abundance of disease vector.

Remote sensing is a new tool that is used in the landscape ecological research. This tool is found to be helpful in enhancing the pattern of recognition. Remote sensing technique is used for analyzing the properties of landscapes of the area according to the state- wise and local area wise by using different technologies like Landsat.

These satellites are used for monitoring the properties so as to maintain and inform about the disasters and changes that are happing naturally as well as manually.

Remote sensing has many applications like, it is used for urban mapping which includes urban and sub- urban areas landscapes, trees, water, etc. secondly it can be used for the hydrological purpose is to quantify the rate of energy, amount of energy and water fluxes in the present environment. Likewise, remote sensing methods can be done for acquiring data related landscapes, water quality, air quality, flood mapping and health- related issues.

Remote sensing is the most widely used method nowadays, even it is being used from the 1970s when there was not much knowledge about this method and not much -developed resources.

REFERENCES

1. 2001. *Remote Sensing Applications At The State And Local Level*. [ebook] Available at: <https://sedac.ciesin.columbia.edu/remote_sens/RemoteSensing.pdf> [Accessed 16 April 2020].
2. Goetz, S., Prince, S. and Small, J., 2000. Advances in satellite remote sensing of environmental variables for epidemiological applications. *Remote Sensing and Geographical Information Systems in Epidemiology*, [online] 47, pp.289-307. Available at: <https://www.sciencedirect.com/science/article/abs/pii/S0065308X00470120> [Accessed 16 April 2020].
3. Melesse, A., Weng, Q., Thenkabail, P. and Senay, G., 2007. Remote Sensing Sensors and Applications in Environmental Resources Mapping and Modelling. *Sensors*, [online] 7(12), pp.3209-3241. Available at: <https://www.ncbi.nlm.nih.gov/pmc/articles/PMC3841891/> [Accessed 16 April 2020].
4. Melesse, A., Weng, Q., Thenkabail, P. and Senay, G., 2007. *Remote Sensing Sensors And Applications In Environmental Resources Mapping And Modelling*. [ebook] Sensors. Available at: <https://www.mdpi.org/sensors/papers/s7123209.pdf> [Accessed 16 April 2020].
5. Rao, D., n.d. *Applications Of Remote Sensing And Geographical Indication System In Land Resources Management*. [ebook] Meghalaya. Available at: <http://www.kiran.nic.in/pdf/publications/Geographical_Indication.pdf> [Accessed 16 April 2020].
6. Spitzer, D., 1986. On applications of remote sensing for environmental monitoring. *Environmental Monitoring and Assessment*, [online] 7(3), pp.263-271. Available at: <https://link.springer.com/article/10.1007/BF00418019> [Accessed 16 April 2020].
7. UKEssays.com. n.d. *Applications Of Remote Sensing Environmental Sciences Essay*. [online] Available at: <https://www.ukessays.com/essays/environmental-sciences/applications-of-remote-sensing-environmental-sciences-essay.php> [Accessed 16 April 2020].
8. Zimmermann, N., Washington-Allen, R., Ramsey, R., Schaepman, M., Mathys, L., Kötz, B., Kneubühlerx, M. and Edwards, T., 2007. Modern Remote Sensing for Environmental Monitoring of Landscape States and Trajectories. *Landscape Series*, [online] pp.65-91. Available at: <https://www.researchgate.net/publication/226804486_Modern_Remote_Sensing_for_Environmental_Monitoring_of_Landscape_States_and_Trajectories> [Accessed 16 April 2020].

9. Adams et al., 1995
10. Aguiar et al., 1999
11. Allen et al., 1998
12. Allen, 1998
13. Anderson et al., 1976
14. Asner and Lobell, 2000
15. Bastiaanssen et al, 1998
16. Carson and Ripley, 1997
17. Chavez et al., 1977
18. Cochrane and Souza, 1998
19. Doorenbos and Pruitt, 1977
20. Florinsky, 2000
21. Kulagina et al., 1995
22. Lee and Lathrop, 2005
23. Lillesand, 2000
24. Lu and Weng ,2006a 2006b
25. Madhavan et al., 2001
26. Mather, 1999
27. McGwire et al., 2000
28. Melesse et al., 2006a, 2006b
29. Melesse, 2004a
30. Meyer et al., 1993
31. Oke, 1982
32. Owen et al., 1998
33. Phinn et al., 2002
34. Powell et al., 2007
35. Rashed et al., 2005
36. Ridd, 1995
37. Senay and Verdin, 2003
38. Senay et al., 2007b
39. Setiawan et al., 2006
40. Small, 2001
41. Smith, 1992

42. Verdin and Klaver, 2002
43. Wang, 1990
44. Ward et al., 2000
45. Weng and Quattrochi, 2006
46. Weng et al., 2004
47. Wu and Muarry, 2003
48. Wu et al., 2005
49. Xie and Arkin, 1997

Chapter 3 GIS Applications for Environmental Monitoring

CONTENTS

This chapter which is named GIS Applications for Environmental Monitoring basically focuses on the GIS technology which can be used in environmental management as well as analysis or monitoring. The first half of the chapter, describes the GIS use in environmental management and environmental modelling and also policy- making.

Furthermore, the chapter puts lights on the role of GIS in environmental studies and what monitoring systems are in general terms. In the second half, the chapter provides insights into the role played by GIS in the management and monitoring of the environment or even resources.

GIS is considered as an analysis tool for the purpose of monitoring data and this has also been covered in one of the topics in the chapter. In the end, the chapter provides knowledge related to the advantages and disadvantages of the GIS while using it in environmental management or modelling.

3.1. INTRODUCTION

Whenever GIS (Geographic Information System) is considered in the environmental modeling applications, then it is said that GIS software is very much helpful in comparison to a programming language.

However, unfortunately, the typical environmental GIS users are much more focused on the issues related to the environment and also, rarely they have the additional technical knowledge which is necessary for the development of new GIS applications.

Furthermore, this causes certain consequences like most of the domain experts are forced to work with GIS which has pre-defined underlying data structures and 'standard' interfaces.

Each type of system is distinct and so, it presents its own demands to the designer. Considering it all together, these requirements may form a 'requirements matrix'. There are some characteristic examples of applications that will be given along with a closer look at the demands that the individual applications make.

First of all, it is useful while distinguishing various types of environmental monitoring and assessment systems. In the monitoring systems, the main focus is put on the data collection, pre-processing and quality control. The analysis systems tend to focus on the usage of the tools in order to manipulate and model data. On the flipped side, the information systems are more concerned with the storage and management of data (Ji and Mitchell, 1995).

Most of the time, the environmental managers are working with spatial data (that is concerned with a specific location on the surface of Earth), and also, with the spatio-temporal data (that is related to a specific location and a specified time).

There are times when the time is referred to as the 'fourth dimension'. The first two dimensions are the location which is represented by an x- and y-coordinate and altitude that is represented by height or z-coordinate.

For the management of the environment, spatio-temporal data are considered to be useful for a number of reasons. Most of the time, the environmental managers make spatial decisions and this in turn, affect a specific portion of a study area.

There can be examples also relating to this fact. The farmers can get paid an agri-environmental subsidy so that they do not use pesticides or fertilizers in water-meadows, but they can feel free in order to apply the agro-chemicals elsewhere on their land.

There can be restrictions put on the tourism activities and new construction within a conservation area, such as a Site of Special Scientific Interest. But outside of such areas, the construction and tourism can grow without even being checked. Environmental monitoring is required to take place so as to evaluate the site-specific management decisions.

This entails collating data for a baseline period and for even one or more periods subsequently. It is important to be monitoring the effect that the decisions have on the environment, in order to be truly adaptive in the management.

In addition to this, there are a number of environmental processes that are inherently dynamic. While underlying geology and landforms remain more or less not so changed considering the period from one year to the next, in such period land cover, climate and species distributions can change rapidly. The environment is changing rapidly so, it needs to be adequately reflected in the spatial databases.

There are issues confronting today considering the process of human evolution. These issues are safeguarding the natural environment along with the maintenance of the good quality of life. While performing the developmental activities, the assimilative capacities of the environmental components such as air, water and land to various pollutions are considered rarely.

There are some developmental activities that are haphazard and erratic which can lead to overuse, congestion and even incompatible land – use as well as poor living conditions. Therefore, the problems related to environmental pollution are becoming a popular topic for the high – risk environment. There is an important role played by GIS in order to analyze and formulate quick mitigation plans for high- risk environments. GIS is considered to be one of the important tools in the environmental data framework for the purpose of data validation, digital data transfer standards, data analysis and data retrieval or dissemination. This can serve as the ultimate communication of environmental information to the public as well as the policymakers because of the fact that it is the technical basis for the multimedia approach while considering the environmental decision – making. The evolution of the spatial data standards, the Internet and the next generation GIS has taken lace which allows all kinds of users to access the information related to the environment in its proper spatial terms.

3.2. MONITORING SYSTEMS IN GENERAL

There are a number of different types of monitoring systems. Among them, there are many which automate the process of data collection and pre – processing. This is a task that is hidden most of the time from the ordinary user. There are a number of different ways in which the implementation of the monitoring can be done depending on the type of system which is under study.

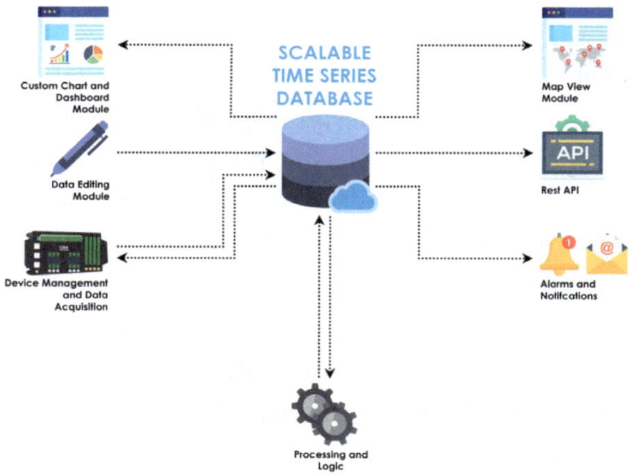

Figure 3.1. Environmental Monitoring System.

Source: Image by Wikimedia Commons

While designing an environmental monitoring system, there are a number of factors which should be considered and they include:

- the acceptable time delay which likely ranges from true real - time monitoring to manual sampling along with the following laboratory analysis;
- most of the time, there is often a distinction that takes place in between manual as well as fully automated data sampling systems; and
- The requirement for controlling the quality which can range from showing pure raw data to a complete quality assurance/quality control (QA/QC) procedure.

There are certain steps which need to follow, it doesn't matter what method is being used. These steps include:

- a communication task that takes data from a sensor (or from any other information source) and communicates it to a receiving monitoring system;
- pre-processing the data by using steps including calibration, checking, and formatting;
- storing the data in some kind of database;
- Presenting the data in a relevant form to users.

The Hiltherto monitoring process and monitoring systems are considered to be rather mechanical. Monitoring may have been done along with the assistance of one or more specialized standard systems like the supervisory control and data acquisition systems (SCADA) and this depends on the specific task.

There is a reason behind the introduction of GIS into the pure monitoring process which is connected most of the time to a user – interface requirement. At this stage, while introducing the GIS, it is possible to use the mapping functions in order to display objects like the measurement station. Moreover, it can be of great value that is to keep the entire system within the framework of the GIS, especially, when the spatial analysis is involved in the required data processing tasks.

3.3. GIS IN ENVIRONMENTAL MANAGEMENT

- Environmental management is said to be not similar to physical geography, environmental social science or environmental science, despite the fact that it contains elements of each discipline. There are disciplines such as geography and environmental science, which use GIS as a tool most of the time. These disciplines provide useful information or insight to the managers. But they are not in "management" themselves even they tend to study the physical processes that impinge on people.

- It is said that decisions precede actions, and therefore, most of the people would consider environmental decision making as an integral part of environmental management. Despite this, there are purists who would argue that there are decisions that are not followed by actions; such decisions do not actually create management.
- Environmental management is fundamentally about "doing". This is specifically concerned with the intervention in the relationship that is possessed by different stakeholders and environmental resources or processes. This works either through the modification of the environment or through adjustment of the people or their expectations. This is exactly where GIS should be put to work.
- Anyone who makes a decision or tends to undertake an action which represents interference in the relationship that is possessed by the people and environmental resources or processes can be referred to as an "environmental manager".
- In addition to this, GIS can also be used in order to provide help while understanding the processes and procedures of environmental management. These procedures or processes are entirely different from understanding the processes of physical geography or environmental science. This is said to be not so similar to the actual management of the environment.

When GIS is required to play a role in environmental management, then the GIS functions is required to be applied to data in such a way that the environmental managers have been able to reach better decisions as well as they become able to undertake better actions. Here 'better' term is defined in broader terms of attributes which include effectiveness, social and political acceptability, sustainability and equity.

There are a number of technical processes which are involved in the application of GIS principles in the environmental management practice:

- Data acquisition and management: data conversion, survey, data quality control, data structure, metadata.

- Data integration (multiple layers approaches) and visualization (that is also referred to as 'mapping') – this includes the combination of variables into representational as well as diagnostic indices.
- Multi criteria approach (synthesizing and relating data sets).
- Spatial and temporal interpolation as well as extrapolation.
- Analysis (this includes area norms and residuals, trends and rates of change).
- Spatial and temporal modeling: there can be an example of spatial modeling and that is predicting the way in which the soil chemistry will seem to be while analyzing data on plant cover. Temporal modeling would be considering the erosion on a steep hillside. This has been caused because of overgrazing of the plant cover and further, predicting the way in which it will develop in the future provided that various different grazing use scenarios.
- Decision support (this includes the handling of fuzzy and uncertain elements).

There is an order in which the functions are placed in the list and there can be debates on this. The only intention is to represent a spectrum from the real world to the management representation and from the simple to complex representation. Nevertheless, despite the fact that the details are being discussed, the entire pattern is robust and furthermore suggests a powerful range of roles played by GIS in environmental management.

These types of roles can be considered to be applying standard GIS functions to a significant extent to environmental management data. In a practical context, the environmental management GIS contains its own specified flavor which is based on the nature of data and the nature of management processes in which the application of GIS is done.

Also, it is said to be important to appreciate the extent to which the GIS application is done in environmental management which are embedded practically in social as well as political processes.

Environmental management is not said to be an objective process. Furthermore, it follows that the application of GIS in environmental management is inevitably and most of the time quite properly subjective, it does not matter how crisp and objective the actual functions GIS might seem to be.

Considering this, it is much clear that there are a number of possible approaches to GIS in environmental management. For this particular purpose, there is a chosen order which is based loosely on the hierarchy of functions which have been mentioned and on the proposed definition of environmental management.

3.4. GIS AND ENVIRONMENTAL MODELLING: STATE OF THE ART

3.4.1. GIS

GIS are said to be computer – based tools which is used to capture, manipulate, process and further, display the spatial or geo-referenced data. These data contain both geometry data (which includes coordinates and topological information) and attribute data, which means the information that describes the properties of geometrical objects such as points, areas and lines.

GIS is quite common. They are generally accepted while performing the survey and mapping, geography and urban and regional planning, cartography (Smyrnew, 1990; Scholten and Stillwell, 1990) and also, the land resources assessment, along with conferences, journals, and monographs while documenting an established field (Burrough, 1986).

In other sense, for environmental applications, (which can include land management) there is a considerable tradition in the field. An example can be of Canada which is done under the header of land modeling in the Lands Directorate of Environment Canada (Gelinas, Bond, and Smit, 1988).

Also, there are a number of major initiatives in order to build up or integrate the geographical and environmental databases in a number of different countries worldwide (Kessell, 1990) and in most of the European countries (van Est and de Vroege, 1985; Jackson, James, and Stevens, 1988; Sucksdorff, Lemma , and Keisteri, 1989) that can be at the European Community level (Wiggins et al., 1986) or at the global level that comes under the UN framework with systems such as GRID (Witt, 1989) or GEMS (Gwynne, 1988).

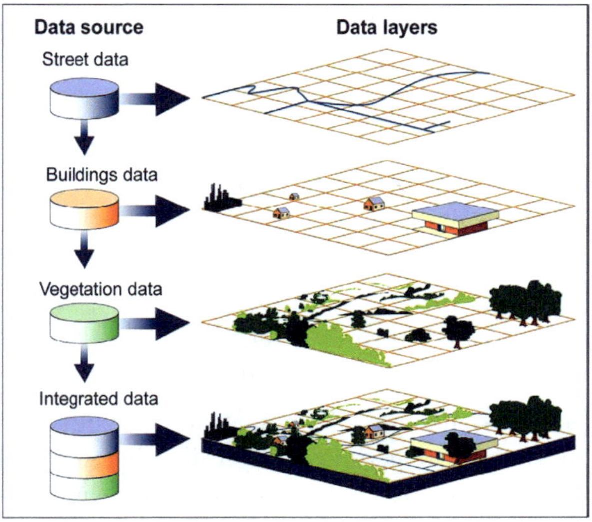

Figure 3.2. Visual representation of themes used in GIS.

Source: Image by Flickr

Despite the fact that there are a number of these systems having explicit environmental components and functions, usually, they are not integrated along with any modeling capabilities in context to simulation models such as models of population development, transport or process and fate models, etc.

However, the idea of such integration is quite obvious and is discussed frequently (Granger, 1989; Tikunov, 1989; Fedra, 1990b; Lam and Swayne, 1991). There are much recent overview papers which include Jackson, James, and Stevens (1988), or Woodcock, Moffat, (1990); Sham, and Shaw (1990); Bishop, Hull, and Bruce (1991) that are based on the environmental GIS and information technology for environmental applications.

In addition to all this, there is an increasing emphasis placed on remote sensing data as a valuable source of environmental data (Woodcock and Strahler, 1983; Welch, Madden Remillard, and Slack, 1988). There is one more interesting line of development that is the integration of GIS and expert systems (Maidment and Djokic, 1990; Lam and Swayne, 1991; Davis and Nanninga, 1985; Davis et al. 1991).

This further aims at a more flexible as well as complex analysis of maps and map overlays that is based on the rules and logical inference. In the alternative context, GIS can provide spatial data to assessment systems that are rule based (Fedra and Winkelbauer, 1991). With this, the related fact is the explicit use of GIS as decision support systems (Fedra and Reitsma, 1990; Parent and Church, 1989).

In spatial databases, when the dynamics or temporality is added then it gets them closer to spatially distributed dynamic models (Armstrong, 1988; Langran, 1988). When the video technology is added for fast animation, then it becomes another feature for better visualization and presentation (Gimblett and Itami, 1988).

There are a number of frequently discussed issues and one of them is the extension of traditional 2D GIS into full 3D systems (Turner, 1991).

In technical terms, the notable trends are moving from proprietary systems to open systems, experiments with distributed systems (Seaborn, 1988; Ferreira and Menendez, 1988) and embracing window environments (Gardels, 1988).

3.4.2. Environmental Modelling

There are considerable history and developments behind the environmental modeling. A larger number of analytical approaches have been applied to biological as well as ecological problems, according to Lotka (1924) and further, the fields such as hydrology can be considered to be more than a hundred years of modeling history. The numerical simulation models have become feasible along with the advent of digital computers.

There has been the development of the early linear models (Patten, 1971), much more complex multi-compartment models such as CLEANER and MS. CLEANER (Park et al. , 1979) and applications of system dynamics to ecological problems (Forrester, 1932).

There have been discussions on limnology, plant sociology and oceanography concepts which include patchiness, however, that are not related to a fixed coordinate system that is used in GIS. When there are more powerful computers then it allows more complex models to be developed and run.

One straightforward way of improving models is spatial distribution and increased dimensionality and resolution. The spatially distributed models can do interactions with the GIS.

3.4.3. Modelling in policy-making

The development of models is done for a particular purpose and scientific research as an end in itself. A better understanding is noble but still is increasingly insufficient to get the funding for research.

The direct responsiveness to actual as well as perceived needs of the society is also important. There was a considerable time at which the discussion is being done on the modeling for decision support or model – based decision support systems considering the environmental problems (Holcomb Research Institute, 1976; Bach mat et al., 1980; Andelin and Niblock, 1982; Loucks, Kindler, and Fedra, 1985; de Wispelaere, Schiermeier, and Gillani, 1986; Labadie et al., 1989; Fedra and Reitsma, 1990; Fedra, 1991).

The integrated environmental models with GIS plays an important role and would largely be in their ability so as to communicate effectively, be using apps as a well – understood as well as an accepted form of information display and furthermore, generating format that is widely accepted and familiar for a shared information basis.

3.5. GIS IN ENVIRONMENTAL STUDIES

GIS is said to be a powerful tool for performing environmental data analysis and planning. In a digital mapping environment, spatial information (data) is stored by GIS. In order to view spatial information and relationships, a digital base map can be overlaid with the data or some other layers related to the information onto a map.

Furthermore, the GIS allows better viewing as well as understanding the physical features and the relationships which tend to influence in a provided critical environmental condition. There are factors such as the steepness of slopes, aspects and vegetation which can be considered and overlaid in order to determine the number of different environmental parameters and impact analysis.

Figure 3.3. Students of environmental studies.

Source: Image by Wikimedia Commons

Also, GIS can help in displaying and analyzing aerial photos. Digital information can rely on photographs in order to provide environmental data analysis having much more familiar views of landscapes and further related data. GIS can give a quick, comparative view of hazards (which include highly prone areas) and risks (like the areas of high risks which can occur) and some areas that are to be safeguarded.

When data analysis is done, then the GIS help in the process of planning and managing environmental hazards and risks. The assessment of hazards and risk becomes the basis for planning decisions and for the mitigation activities and this is done in order to plan and monitor the environmental problems.

There are activities in the environmental assessment, monitoring and mitigation which are supported by GIS. The GIS can also be used in order to generate environmental models. There are some application areas where the GIS can be implemented for the purpose of effective planning and management.

GIS applicable areas are given below:

- Emergency Services like Fire Prevention
- Wild Land Analysis
- Hazard Mitigation and Future planning
- Disaster Management
- Air pollution & control
- Forest Fires Management
- Waste Water Management
- Managing Natural Resources
- Oil Spills and its remedial actions
- Coal Mine Fires
- Sea Water – Freshwater interface Studies

3.5.1. GIS in Field

In laboratories, apart from data analysis, GIS can also be helpful in environmental data analysis in the field. GIS is said to be much flexible so as to work in the field in order to give the location of devastation and amount of devastation. There are some examples in the field where the application of GIS is done and these examples include:

- It helps in retrieving previous inspection records in order to compare with the existing environmental conditions.
- An environmental inspector can use GIS in the field which can rapidly map waste storage sites; describing the volume, content and state of waste containers.
- When the GIS is integrated with a global positioning receiver, then a field crew can use GIS so as to ground- truth satellite imagery in oil spill mapping accurately and its effects on the surrounding ecosystem.
- It helps in viewing the environmental data in relation to adjacent geographic features which include waterways, neighborhoods, or

other sensitive areas such as high-risk zones for landslides, water pollution, etc.

3.5.2. Constraints using GIS

There are some of the intricate problems that occur while implementing GIS in environmental studies. The problems which are posed with this scenario include:

- Meager skilled Manpower in the government planning and development departments;
- Non- availability of properly spatial data;
- GIS software being more costly;
- Lack of proper infrastructure with the Government bodies.

There are some probably solutions to these problems which include:

- Supplementing the Environmental planning division with adequate funds
- Availability of map data in a centralized facility;
- Increasing the infrastructure facilities to cope up with the latest technologies; and
- Awareness and increasing the skills proficiency in GIS in government and private sector.

3.6. GIS AS AN ANALYSIS TOOL FOR MONITORING DATA

There are times when the use of GIS for the purpose of environmental monitoring is restricted to the presentation of data and the creation of maps. The inherent powers of GIS are used as a tool for the purpose of data integration and analysis which are being under-utilized most of the time.

Monitoring systems often tend to stop the data analysis process when the adequate quality control of the incoming data has been performed and furthermore, it tends to prevent environmental authorities from using GIS in order to perform the investigation of the causes of pollution.

Figure 3.4. A woman analyzing monitoring data.

Source: Image by Air Force Civil Engineer Center

There can be gradual changes in this which will take place along with the creation of more and more databases having spatial information.

This will happen so that the information on the industrial development can readily be combined on a more ad hoc basis in comparison to the case present at that moment.

There is a very important method for strengthening the general awareness of environmental conditions and that is the ability to aggregate data from a number of different inputs and further present them in the form of a map.

Also, GIS is said to be a valuable tool considering the process of decision – making. There are some who use GIS in order to present environmental data and consider them as maps or graphs.

Those are required to be aware of the issues that are related to the classification, symbolization and visualization. The reason behind this is that they can affect the conclusions that are drawn by the people about the data.

3.7. THE ROLE OF GIS IN ENVIRONMENTAL MONITORING

GIS plays a very important role in environmental monitoring. Ideally, GIS is considered as a tool for presenting the data which is derived from the distributed measurement stations such as field-based water quality sensors. However, there are much GIS which has some severe shortcoming when it comes to dealing with the typical data that is obtained from such measurements such as time -series data.

In other words, it can be said that the time series is long consecutive runs of data which include the temperature measured every half hour at a specific point.

It is important to be extended or even auxiliary application interfaced, because of the fact that most standard GIS software packages do not contain adequate tools for the purpose of handling the temporal data.

Abel et al (1994); Federa and Kubat (1992); Goodchild et al (1992); Maguire (1995); Steyaert and Goodchild (1994) have discussed the approaches in order to integrate GIS and external software systems. Routinely, the time series data are collected for performing the environmental monitoring and which most of the time have two considerable analytical and display problems.

The very first point is that they typically relate to points when the data is based on aerial units and would be more useful most of the time (which is a problem of spatial object type).

The second point is that the temporal sampling incidence is too frequent most of the time which further leads to large volumes of data. GIS provides relatively simple means for dealing with each of these problems.

A measurement point can be permanently assigned to a representative area, with regards to spatial resolution, by using a simple point to point transform, which can include those that are based on Thiessen tessellations.

There can be one other option also and that is to use standard GIS gridding and contouring facilities in order to interpolate between the point measurements across the entire study area. The standard database selection capabilities can be used with regards to temporal resolution in order to extract data that pertains to a time value corresponding to some other data to which the comparison of the same can be done.

There can be an alternative way also, and that is the standard descriptive statistical functions that can be used in order to reduce the data volumes, for example, calculating the average or maximum values for the characteristic periods. From these approaches, any of them can be used alone or even in combination.

GIS is only used at user workstations, while there is a major part of the system which is dedicated to some other tasks. In most of the cases, this distribution is reflected in the corresponding implementation and running costs.

Only a minor part is related to the programming of the analysis and display functions that are based on GIS because of the fact that at least 90 percent of the cost that is included in the setup of environmental monitoring systems is related to the measurement program, processing of data and the quality control.

Most of the time, there are simple data analysis functions that are provided by standard GIS software packages. These are not sufficient in order to deal with the data monitoring requirements of environmental problems. There are a number of reasons for this, and most of them are:

- The economic aspects of a project which do not allow sufficient data to be collected in order to feed the tools being used in the analysis;
- A number of different parameters which are interconnected and the required data cannot be measured directly; and
- The characteristics of the data tend to make simple extrapolation inappropriate.

Considering such cases, it can be said that when the models are much more complex than it is required to reflect the expected environmental situation. In such a situation, the task of environmental monitoring goes further which can be achieved within a standard GIS in real.

3.8. MANAGING THE ENVIRONMENT USING GIS

For all of the people, in the current scenario, the most important concern is to protect the environment where they live and where they breathe in.

There are climate change issues that are taking place and furthermore, they are creating havoc with erratic weather patterns. These patterns are affecting everything be it crop production of untimely melting of ice glaciers.

It is important to worry and take immediate action. The world has also geared up in order to take corrective actions, but it is important for people also to do more, and GIS can help us in achieving the same. GIS is definitely a powerful tool. Furthermore, it is making every sector able to perform something better and the environment is no exception.

3.8.1. How GIS can help?

The environment is degrading and human activities and global warming is contributing much to that. This causes a decrease in the glacier area, increase in the glacial like size, changes in the land use and land cover, unprecedented rainfall, flood and glacial lake outburst floods, forest degradation, landslides and shortfalls in the production of the agricultural crops. There are a number of other problems also which are brought on by environmental changes.

These are the issues that need to be monitored and supervised timely. It is important to have valuable information and data for the purpose of effective monitoring of the environment and an improved understand related to the same. The information or the data can be extracted by applying geospatial technologies which include remote sensing and GIS.

GIS can be further used for the purpose of environmental data analysis and planning much more effectively. Also, it allows a better view and understanding of the physical features and relationships.

These relationships influence a provided critical environmental condition. There are factors that can be viewed and overlaid in order to determine the number of different environmental parameters and impact analysis. These factors include steepness of slopes, aspects and vegetation.

GIS also helps in displaying and analyzing aerial photographs. The digital information can be overlaid on photographs.

This further provides environmental data analysts with views of landscapes that are much familiar and some other related data also. A quick as well as comparative view of hazards of the highly prone areas, is provided by the GIS and it also helps in providing details about risks in the areas having high risk and areas that are to be safeguarded.

Along with the completion of data analysis, GIS can help in effective planning as well as managing the hazards and risks related to the environment.

The assessment of hazards and risks tends to become the foundation for the planning decisions and for mitigation activities in order to plan as well as monitor the environmental problems.

There are activities in the environmental assessment, monitoring and mitigation which are supported by GIS and GIS can also be used in order to generate environmental models.

GIS can help in hazard mitigation and future planning, disaster management, air pollution and control, managing natural resources, forest fires management, oil spills and its remedial actions and wastewater management and many more.

3.8.2. GIS in Disaster Management

GIS can help in mitigating the risks of a disaster to a great extent whether it is modeling which can be done through early warning systems or the usage of decision support systems in order to understand the disaster that is going to affect or is now affecting which region the most.

Better preparations can be done by using GIS. There can be efforts that can be directed and further, the response can be faster. Furthermore, GIS tends to enable the response teams in order to gain situational awareness, engage with the public and also, understand the effect of the event. Recovery becomes easier and faster because of the fact that GIS results in better identification of the affected areas and people.

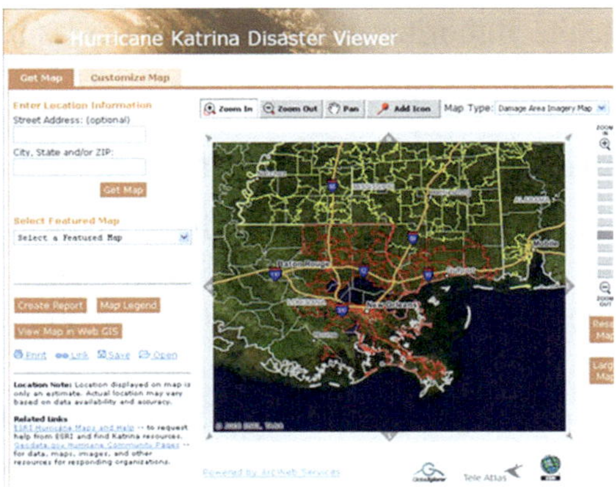

Figure 3.5. GIS is very useful in disaster management.

Source: Image by Flickr

An example can be taken from India. India has made great strides in disaster alert systems in the recent past. These disaster alert systems include regional tsunami warnings, heavy rainfall or flood alert systems, or the cyclone alert system.

The Indian Tsunami Early Warning Centre is based in Hyderabad and has been successful while delivering accurate alerts. Through this, timely predictions are done and due to this, the preparations have been better, further leading to timely evacuations and therefore, no loss of lives.

3.8.3. GIS in Air Quality Monitoring

It is important to monitor air quality and has become a critical part of healthy living. GIS plays a very important role in air quality monitoring also. There is a GIS integrated platform by leveraging sensors and IoT for the purpose of air quality monitoring, analytics and planning. These can accurately predict the PM levels in a number of various areas within any city.

GIS also helps in providing information like which areas are the most hazardous or most dangerous for every person, more particularly for the patients suffering from asthma. This type of analysis can provide help to field officers so that they can take corrective action on time so as to improve the air quality.

Citizen engagement is becoming an important part of such kind of applications also. Citizens can use their mobile phones and some mobile apps and can make the authorities aware of which areas require immediate actions.

3.8.4. GIS in Forest Fire Management

There can be a huge loss of flora and fauna which can be caused by wildfire. The very first strategy in order to defend the forests against the wildfire is to avoid it. GIS has the ability and has proved its potential in the management of forest fire.

There are a number of different applications of GIS in forest fire management. Among these, there are ones which are most important and that are the hazard map production, resource management and forest fire simulation.

Simulation plays a main role in the management of forest fire. There are various information layers that are used by GIS. These include DEM that is the Digital Elevation Model and the index of flammability along with a number of different models for the management of forest fire.

Figure 3.6. GIS – a framework for identifying and managing forest fire.

Source: Image by National Park Service

3.8.5. GIS in Managing Natural Resources

Identification of the impact of human behavior on natural resources can be done by GIS and this further leads to the more effective utilization of the same. Remote sensing, aerial photography or satellite imagery can be used in order to collect the data related to natural resources and further, they are mapped by using the GIS technology.

Majorly, GIS can be applied in the management of natural resources and that is in confronting environmental issues such as landslide, flood, drought, soil erosions and earthquake, etc. also, it tends to address the current problems related to habitat loss, climate change, pollution, population growth, etc. GIS also provides information related to the land area changes between the time periods.

The information that is obtained from GIS helps in studying the particular areas and further monitoring can be done in and around those areas. It helps in providing appropriate information related to the environmental condition and policy which include conservation programs. The maps in GIS tend to provide information related to location and the current resources.

In addition to these applications, GIS can effectively help in wastewater management, sewage treatment, oil spilling, etc. The spatial information can result in better outcomes in almost every industry. GIS also provides some invaluable location information which makes the decision – making superior and also, extremely productive.

There are many benefits of GIS and these benefits are only restricted to the capability of humans to innovate and harness. It is important for humans to dedicate themselves to preserving the environment and technologies such as GIS which can prove to be one of the best for the purpose.

3.9. ROLE OF GIS IN IMPROVING ENVIRONMENTAL ASSESSMENT EFFECTIVENESS: THEORY VS. PRACTICE

3.9.1. Overview of the Potential of Using GIS for EIA

The GIS can help in preparing the data for performing analysis more quickly and accurately in comparison to the analysis that normally possible by manual procedures. Let's take an example.

There are a number of procedures that can be carried out relatively easily and these include changes of scale and projection, conversion of the map data into a form of a grid and the changes of the weights of the different variables. Also, the GIS can help in coping with a higher level of details and thus, it increases the accuracy of the data handling process.

GIS can prove to be powerful spatial analysis machines so as to be used in environmental impact assessment. The process of EIA that is environmental impact assessment includes the consideration of a number of different variables as well as phenomena. These have complex interrelationships in most of the time which can further vary in space and time. GIS can also be considered as an important tool while modeling such variables.

GIS can be used for environmental modeling or analysis and there are two main ways that are involved in the same; the GIS is carried out within the GIS or it is carried out by connecting the GIS to a computer model externally. In both cases, the use of GIS can be advantageous.

3.9.2. How Are GIS Being Used In Practice In EIA?

It has been found that the GIS is being used in the whole world for a wider range of EIA uses in comparison to it was anticipated by the authors. In fact, it was found that the GIS is being used by EIA in the current scenario by the EIA practitioners considering all the EIA stages. GIS tends to help in gathering data, impact assessment, spatial modelling and the calculation of impact magnitude for the preliminary EIA stages related to screening and scoping.

GIS is also being used for the purpose of a description of the project where the GIS can help in putting the projects in a geographical manner. The environmental baseline is established which tends to get benefitted from GIS manipulations such as documentation of baseline conditions.

These include hydrology, biophysical inventories (including vegetation, land use, habitat, etc.), archaeological as well as historical resources, soils, topography, land ownership, utilities, roads and many more.

The main stage of EIA is impact identification. This stage can benefit from typical GIS operations such as overlay analysis.

An example can be taken as GIS is being used in order to overlay the pollutant distributions having resource maps or for the purpose of integrating the results of air quality modelling and habitat suitability analysis.

GIS can be applied to the quantitative assessment of the percentage of resource base which is affected by pollutants in order to predict the impact magnitude.

Also, it is possible that the creation of impact magnitude maps is done which are derived from the integration of the results of risk and air quality modelling having other data layers which include soil susceptibility to acidification.

GIS is specifically useful in order to show the spatial distribution of the impact significant and the way in which the various changes along with the different project alternatives in order to assess the impact significance.

Considering the stage of EIA that is of impact mitigation and control, GIS are going to be applied in order to model and map distributions of pollutants. Then, the result can based on the user so as to direct the human power and equipment to the most effective clean – up location. Also, the GIS are being used in order to show the geographical location and the extent of mitigation activities over the period of time.

GIS are being used for public consultation and participants. This is done in order to prepare the presentation material, to explain the project to the public and furthermore, to allow a quick response to the required changes.

Ultimately, for the purpose of monitoring and auditing, GIS are relevant in order to prepare the designs of monitoring programs, to process and store the monitoring data, to compare the actual outcomes along with predicted outcomes and to present the data which shows the variation of the location of pollutants with time.

Furthermore, it can be observed that the application of GIS can be done for the purpose of analysis or modelling for around EIA stages from scoping to monitoring. Using GIS for the purpose of analysis or modelling for different EIA stages proves that using GIS in EIA studies is not limited to mapping the production and preparing the report.

Nevertheless, there was many exceptions to the use of GIS in EIA and that is the complete absence of any mention by EIA practitioners regarding the use of GIS in the public participation in the areas for which data is being prepared and analysed or modelling is being done.

GIS are only being used for the purpose of communication of the final results to the public and furthermore, no attempt is being made for the collection and integration of views of the public in the EIA process through GIS.

There is a small but also an important body of literature that tends to explore the potential of using GIS for the incorporation of the point of view of public prior to planning decisions are taken and this literature includes Faber et al. 1996.

3.9.3. Advantages and Benefits in using GIS for EIA

According to the EIA practitioners, the most important thing was the ability of GIS in order to perform spatial analysis and modeling which include spatial analysis of resources and pollution, spatial queries, constraint analysis, overlay operations and much more.

GIS can make some type of analysis much easier to perform. And also, GIS can also allow some new analysis or modeling which would be impossible or practical to perform otherwise. There is one most frequently quoted advantage about the usefulness of GIS and that is that they provide the clearer presentation of results.

These results are related to the capability of GIS in order to display spatial data. That seems to be useful from the traditional digital thematic mapping to more advanced as well as complex 3-D visualization.

GIS has the power to store, manage and organize the spatial data and this is useful even for the very large data sets. They have the ability to integrate and manipulate various kinds of spatial information which were previously unrelated and this is considered to be the next most preferred advantages that are suggested by the EIA professionals.

There is one more important advantage and that is the easy way in which the GIS allows updating as well as changing of information and further, obtaining new results in order to change the conditions.

These conditions are important for the generation of alternative scenarios and so are crucial for studies related to EIA. There were other advantages also which were mentioned, even though less frequently, which include flexibility and speed, accuracy, precision and reliability.

For EIA, these findings corroborate the descriptions of the potential benefits that are of using GIS when the research literature is considered. There were some by Eedy in the year 1995. As per the suggestions provided by Eedy, the key reasons for these systems being served as valuable tools in the development of impact assessments are mentioned below:

- They have the ability to score large, multidisciplinary data sets (which include the storage and manipulation of three-dimensional files);
- There is the possibility of systematic updating of the database, along with the consequence of being able to use similar data for more than one project or for various mathematical models;
- They have the power in order to identify complex interrelationships between environmental characteristics (which include evaluation of changes over time); and
- They have good visual display capabilities that lend themselves specifically well in order to explain the development plans or alternatives to the public.

3.9.4. Key Problems or Disadvantages in Using GIS for EIA

Contrary to the advantages that are mentioned above, there are many problems also which are related to GIS technology. These problems are preventing the GIS to be used more widely or to be used to their full potential.

The disadvantage includes the long time that is required by GIS to set up, to build the database and also, to input as well as process the appropriate data. These are considered to be key constraints and are probably the most important factors in order to prevent the more widespread use of GIS in environmental consultancies.

For this, the reasons are two – fold. On one side, there are many environmental assessment projects which have strict deadlines and therefore, there can be less time or not much enough time for the preparation of data and input into the GIS.

One the flipped side, even if there is enough time for the preparation of data, but there can be the case when the data will not be required again. Then the effort of creating such a particular database cannot be considered worth. In other words, it can be said that there is no point in building a sophisticated database if it is going to be used only once.

While using GIS for EIA, there is another most important constraint. This is the cost of initial start- up for hardware and software acquisition. The most important one is the cost that is incurred for the data collection and conversion to digital form.

As per the views of Huxhold and Levinsohn (1995),

"*the cost of converting data to digital form exceeds by far the cost of any other component of the system (between 50 and 80% of the total cost of the project)*."

The cost implications are related to this and that is for the time it takes in order to make the system operational. There are some other disadvantages also which are pointed out and these include the lack of data in digital format and the increased requirements in context to training and technical expertise.

Data error and accuracy considerations are considered as the key problems related to GIS. Building as well as operating the GIS can require staff to have some of the specialized skills and the new staff have to be hired or the existing staff will need extra education and training.

While using the GIS for the environmental modeling, there are two problematic issues and these include the difficulty in knowing the accuracy of the data that is inputted into the system and then tracking the error propagation because of the GIS manipulations (Goodchild et al. 1996 and Thapa and Bossler 1992).

There exists a common problem and that is of data gaps. This is difficult to solve. In this issue, the researchers have to collect more data that might not be viable or they have to aggregate upward and risk more data gaps. This issue is the one that can result in a misunderstanding by users which can include how to interpret the data interpolation or extrapolation.

The possibility of errors related to users has also been considered as a disadvantage while using the GIS techniques which are computer -based instead of manual methods that are more conventional.

Moreover, this contains the lack of critical thought related to inputs or choices which entered into a GIS. In EIA, accuracy issues and data quality checks are considered to be specifically important, especially keeping in mind the environmental statements that can be used as legal documents. (Epstein 1991)

There are some environmental consultancies which described the weak analytical capabilities of GIS for specific purposes as a disadvantage of the systems. According to Kidner (1996), there is some disappointment with results while using the GIS for visual impact assessment.

Partly, the reason behind this was the specific problems related to the view-shed algorithms which are available in current GIS.

There can be one more reason and that is the high expectations of the landscape architects regarding the capabilities of GIS. The landscape architects expected the GIS that they will perform exceptionally better in comparison to any other traditional procedures while considering the expense or cost of GIS.

3.10. CONCLUSION

GIS plays a very important role in every part of everyday life. GIS has grown and modified from a simple numerical representation of geo-data into a self-sustaining visual database and data processor. The emergence of GIS has taken place into a new powerful discipline of Science.

Nowadays, GIS Science provides a fast -growing industry of motivated GIS scientists and users which are present around the globe. These people help the decision- makers in order to take relevant decisions by promptly following the sound logic that exists behind each decision.

GIS technology was and will continue to play a very important role in the environmental system management. The GIS has become the primary repository of the information.

GIS can quickly be accessed as well as viewed whenever required. GIS is now becoming much more sustainable for the emergency operations and also, this technology is integrating tools that allow real-time display of information.

The conclusion that can be drawn from the above chapter is that GIS tends to provide rapid access to information, safety, efficiency and better resource management. GIS technology has the ability to provide critical information in some time so that remedial measures can be taken in no time as effectively as possible.

REFERENCES

1. Ali, K., 2015. *Environmental Application of GIS.* [ebook] p.2. Available at: <https://www.researchgate.net/publication/281701258_Environmental_Application_of_GIS> [Accessed 14 April 2020].
2. Fedra, K., 1994. *GIS AND ENVIRONMENTAL MODELING*. [ebook] p.3. Available at: <http://pure.iiasa.ac.at/id/eprint/3730/1/RR-94-02.pdf> [Accessed 14 April 2020].
3. Generic.wordpress.soton.ac.uk. n.d. *Introduction To Unit 6: Environmental Monitoring With GIS | GEM.* [online] Available at: <http://generic.wordpress.soton.ac.uk/gem/unit-6/> [Accessed 14 April 2020].
4. Generic.wordpress.soton.ac.uk. n.d. *Unit 1: Introducing GIS In Environmental Management | GEM.* [online] Available at: <http://generic.wordpress.soton.ac.uk/gem/unit-1/> [Accessed 14 April 2020].
5. Geospatial World. 2009. *GIS In Environmental Studies - An Overview - Geospatial World.* [online] Available at: <https://www.geospatialworld.net/article/gis-in-environmental-studies-an-overview/> [Accessed 14 April 2020].
6. João, E. and Fonseca, A., 1996. THE ROLE OF GIS IN IMPROVING ENVIRONMENTAL ASSESSMENT EFFECTIVENESS: THEORY VS. PRACTICE. *Impact Assessment*, [online] 14(4), pp.371-387. Available at: <https://www.tandfonline.com/doi/abs/10.1080/07349165.1996.9725913> [Accessed 14 April 2020].
7. Larsen, L., n.d. *GIS In Environmental Monitoring And Assessment.* [ebook] p.3. Available at: <https://www.geos.ed.ac.uk/~gisteac/gis_book_abridged/files/ch71.pdf> [Accessed 14 April 2020].
8. Paul, S., 2018. *Managing The Environment Using GIS.* [online] Geospatial World. Available at: <https://www.geospatialworld.net/blogs/managing-the-environment-using-gis/> [Accessed 14 April 2020].
9. Abel et al, 1994
10. Andelin and Niblock, 1982
11. Armstrong, 1988
12. Bach mat et al., 1980
13. Bishop, Hull, and Bruce, 1991
14. Burrough, 1986

15. Davis and Nanninga, 1985
16. Davis et al. 1991
17. de Wispelaere, Schiermeier, and Gillani, 1986
18. Eedy, 1995
19. Epstein, 1991
20. Federa and Kubat, 1992
21. Fedra and Reitsma, 1990
22. Fedra and Winkelbauer, 1991
23. Fedra, 1991
24. Ferreira and Menendez, 1988
25. Forrester, 1932
26. Gardels, 1988
27. Gelinas, Bond, and Smit, 1988
28. Gimblett and Itami, 1988
29. Goodchild et al. 1996
30. Granger, 1989;
31. Gwynne, 1988
32. Holcomb Research Institute, 1976
33. Huxhold and Levinsohn, 1995
34. Jackson, James, and Stevens, 1988
35. Ji and Mitchell, 1995
36. Kessell, 1990
37. Kidner, 1996
38. Labadie et al., 1989
39. Lam and Swayne, 1991
40. Langran, 1988
41. Lotka, 1924
42. Loucks, Kindler, and Fedra, 1985
43. Maguire, 1995)
44. Maidment and Djokic, 1990
45. Parent and Church, 1989
46. Scholten and Stillwell, 1990

47. Seaborn, 1988
48. Sham, and Shaw, 1990
49. Smyrnew, 1990
50. Steyaert and Goodchild, 1994
51. Sucksdorff, Lemme!a, and Keisteri, 1989
52. Thapa and Bossler, 1992
53. Tikunov, 1989
54. Turner, 1991
55. van Est and de Vroege, 1985
56. Welch, Madden Remillard, and Slack, 1988
57. Wiggins et al., 1986
58. Witt, 1989
59. Woodcock and Strahler, 1983
60. Woodcock, Moffat, 1990

Chapter 4

Remote Sensing and GIS Technology for Monitoring Prediction of Disaster

CONTENTS

This chapter is about different methods that are being used for scanning the earth's conditions or environment which is being disturbed by natural disasters or by manmade disasters. This chapter explains all the possible methods that are used from traditional time till now for having clear imagery about the condition of a particular area which is affected by the disaster.

These disasters can be a natural disasters, which is made naturally. There is a deep explanation about how much effect does this disaster has on the particular area, and using that information there is also further predictions.

For this purpose, different Remote Sensing and GIS methods are used which are explained below.

4.1. INTRODUCTION

In the whole world in the recent past, it has been noticed that whether it is a natural hazard of destruction created by human activities, disaster has become like a big issue of the rising alarm.

Within the global system, it is considered that a natural disaster is an intense event that is responsible for bringing out many major changes in the environment in a short duration of time.

In the outcome, these disasters have led to major injuries or deaths of living organisms. With all this destruction, it also smashed up the expensive properties like communication system, infrastructures and it also brought a loss of natural wealth which includes agricultural land, forest, environment, etc.

They did also have a strong effect on the nation's socio-economic system. It always depends on the nature of the disaster, the impact done on the environment will depend on how rapid or slow effects are, like in case of earthquakes, drought or tsunami respectively.

It also depends on the magnitude and frequency of occurrence according to which disaster may vary in their duration, trigger, scale and necessary actions. As natural disaster arises in many of the parts of the Earth and it is considered that each of the disaster types is basically confined according to the specific regions.

It has been estimated that there are more than 95 percent of the death that happens in the developing countries occur due to natural disasters. These places are specifically exposed to disasters because of the presence of dense- packed population and the poor infrastructures.

These all reasons basically get coupled with the unbalanced landforms and with that they also have continuous exposure that is serving towards weather changes. From the past four decades, it has been noticed that disaster events such as floods, volcanic eruptions, earthquakes, storms, etc., are responsible for creating great havoc to the livelihoods.

They even bought down the socio- economic status of the country which is apart from the damages caused by these disasters to the environment. This disastrous event was responsible for swabbing out the years of urban development by trashing up all the buildings and other infrastructures. These disasters even killed thousands to millions of people (Raheem et al., 2000).

There is one such example that caused an immense loss to the lives and properties are 2011 Tsunami which happened in Japan. Disaster is classified in several ways:

- Human-induced disasters: - induced disaster are basically the natural disasters that are being accelerated/aggravated by human influence;
- Human-made disasters: Human- made disasters are the events which are caused by human activities. For example, industrial chemical accidents, atmospheric pollution, nuclear accidents, oil spills, major armed conflicts, and
- Natural disasters: Natural disasters are called as an event which is caused purely by natural phenomena. They are considered to bring damage to human societies, for example, like volcanic eruptions, earthquakes, hurricanes.

It is believed that the mitigation of natural disaster can be only successful only when there is detailed knowledge which is being obtained about the expected character, magnitude and frequency of the hazardous event that takes place in that area.

There are many types of information that are needed for the management of natural disasters that are considered as an important spatial component. Spatial data are the data that contain geographical components, like maps, satellite imagery, rainfall data, aerial photography, borehole data and GPS data, etc.

The data which is received from these different components are considered to have the data with different projections and coordinate systems. They are also needed to be brought to a common map basis; this is done so because in order to superimpose them.

As now there is access to information gathering and organizing technologies of Geographic information system (GIS) and remote sensing. This has proven that there is usefulness in disaster management.

Remote sensing and GIS basically provide a database. This database is basically the data that depends on the evidence that is being left behind by these disasters that might have occurred before and can be interpreted.

This information is further combined with the other information that helps in reaching the hazard maps, which basically indicates which areas are possibly dangerous.

This zoning of hazard should be on the basis of projects of disaster management and it should also supply the planners and decision -makers with the information, which should be adequate and easily understandable.

Remote sensing data which are in the form of satellite images and aerial photos, they all should allow the readers to map the variability of terrain properties. These terrain properties are such as water, vegetation, and geology which include both time and space.

Satellite images give a synoptic overview and also provide useful information about the environment, this information is used for a wide range of scales, which can be for the entire continent or it can be a detail of a few meters.

Secondly, the satellite can detect the early stages of the events like drought, flood, volcanic eruption, etc., which are considered as anomalies in the time series.

Data that is required for the disaster management is considered to be received from different scientific disciplines and these information's be integrated. In the GIS, data integration is considered as one of the strongest points. Generally, there are certain types of data which is required:

- An element that is being destroyed by some of the disaster events then the data is calculated by the settlements, socioeconomic data, population, infrastructure, etc.
- Disaster phenomena like flood, earthquake, landslide; data are achieved from their location, magnitude, frequency.
- Emergency relief resources data is being achieved by hospitals, warehouses, police stations and fire bridges.
- Data on the environment -related disaster event take place through geology, topology, geomorphology, soils, land use, vegetation and hydrology.

4.2. APPLICATIONS OF REMOTE SENSING IN DISASTER MANAGEMENT

4.2.1. Earthquakes

Earthquakes are believed to occur in the cycles of centuries or decades. There are some scientists of the U.S. Geological Survey (USGS) who have operated seismographic stations for more than 35 years throughout the world.

From the past few years, it has been noticed that the Incorporated Research Institute for Seismology (IRIS – a consortium of more than 90 universities), the USGS upgrades its system into a state-of-the-art Global Seismographic Network (GSN).

GSN is designed in such a way to obtain a high quality of data in the digital form. This data is made in such a way that it can readily be accessed by the users of data worldwide.

There are some stations where the data is being reported to the orbiting satellite. After that, the data is then sent to the internet where the information can be viewed by the people using the World Wide Web (WWW).

The remote sensing technique can also add-up the information that is available by the seismic technique. Generally, it is noticed that the fault which is associated with the earthquakes can be identified on good resolution satellite imagery. But the earthquakes related to the volcano are not that clear (Richards, 1982).

For this purpose, it was believed that the geological maps and land use can help to give the vital pointers that will help in pointing the potential earthquake zones. In this case, the satellite sensors that are active in both the near- infrared and visible infrared spectral band can be helpful.

Although, IRS (IRS-1D Handbook, 1997), SPOT (www.SPOTimage.fr), IKONOS (www.spaceimaging.com), NOAA (www.usgs.gov) and LANDSAT (www.nasa.gov), all of them are collecting the required data.

In the above satellites, LANDSAT imageries are considered to be most popular because it has a long historical data recorded of the satellite and it is also considered as cost -effective.

However, airborne radar or aerial remote sensing is considered to be more effective to define the unconsolidated deposits that are found to be active on fault zones.

This fault zone is considered on which most of the destruction takes place and for identifying the areas where an earthquake can trigger the situation of landslides. But now, with the 1m resolution capacity, the satellite imageries professional is seen to be very hopeful, and they apply the technique of remote sensing in a large amount.

Figure 4.1. An image representing earthquake imagery with 1m resolution.

Source: Image by USAID Impact

In the above figure, an illustration of earthquake imagery is being shown using a 1m resolution image. The image showing extensive damage is caused to the individual building as a result of an earthquake.

4.2.2. Volcanic Eruptions

There are approximately 500 active volcanoes that are present around the globe, in which there are about 100 of which are expected to erupt every year.

Volcano monitoring is considered to be important because of an unexpected awakening of these volcanoes can imperil about thousands of lives over world wide areas.

Remote sensing technique plays an important role by providing all the vital information with only limited fieldwork, which indirectly saves lots of time, money and effort. Whereas thermal infrared (TIR) imagery is capable of capturing the volcanic heat, this provides spatial resolution.

Also, PAN stereo-pair imagery is used for capturing an image of a volcano, as it has 3-D capabilities with moderate resolution. This basically serves the purpose of finding out the evidence related to hazardous activities.

An IR patter is received which shows the geothermal heat in the vicinity of a volcano which indicates the thermal activity. This pattern is displayed by many inactive volcanoes.

There are many volcanoes that are extinct and they may also have been reclassified if during regular monitoring any abnormal high IR emissions are obtained from either the summit craters or by the flanks. Changes in the thermal pattern can be obtained only with the help of periodic IR imageries of high resolution.

There are various sizes and nature of volcanic hazards that require the use of various types of sensor that are present in both satellites and aircraft. It is being noticed that relatively small area is involved with the volcanoes. So, because of this reason, the use of aerial photography should be encouraged for their analysis.

It is considered that panchromatic black and white stereo aerial coverage is suitable for recognizing and mapping the geomorphic evidence about the associated hazards and recent activity. This panchromatic black and white stereo coverage basically range between 1:25,000 to 1:60,000.

Color and color IR photography may be considered to be useful for determining the possible effects of volcanic activity done on nearby vegetation. As there are some disadvantages like low resolution, slow film speed and high cost diminishes much of any advantage they are providing.

The airborne thermal IR is considered as the most valuable tool that is used in surveying the geothermal state of a volcano.

The heat produced within the volcano and its movement is easily amenable to detection, as there is a rapid decrease in the resolution with the increase in the altitude that is about 2m per 1,000m. Therefore, the survey is needed to be made at a low altitude which about 2,000m.

There is an IR pattern of geothermal heat is seen in the vicinity of a volcano which is called as an indication of thermal activity. This type of IR pattern is observed in the case of many inactive volcanoes.Thermal IR bands are used commonly which are resulted from the satellite sensor, which is now seen to be available is considered to have an inadequate spatial and thermal resolution. This is considered to have a significant value that is used for detecting the dynamic changes in the volcanic geothermal activity.

In addition, with the sensing geothermal heat, there is another remote sensing technique that is being used for preparing volcanic hazard zonation map and they are also used in case of mitigation volcanic hazards.

Mitigation technique often requires photo interpretation and topographic maps which includes predicting the path of the potential mudflows or lava flows. They even restrict development in those areas.

4.2.3. Tsunamis

Tsunamis are the water waves or seismic sea waves. These waves are caused by a large-scale and sudden movement of the sea floor. This movement can occur due to landslide, earthquake, volcanic eruption or can be caused by man - made explosions.

Figure 4.2. An image representing two images, one is before Tsunami and another is of after Tsunami

Source: Image by Wikimedia Commons

With the increase in the population and development that is taking along the coastlines, these causes are correspondingly increasing the risk of the tsunami disaster in recent years. Tsunamis are different from earthquake hazards. Tsunamis are believed to cause serious damage up to thousands of kilometers from the causative faults.

It is noticed that once they are generated, they are considered as nearly imperceptible in the mid -ocean. This so because their surface height is less than a meter in the mid- ocean. Tsunami is seen to travel at incredible speeds which are approximately 900 km/hr, whereas, the distance that is noticed between the waves crests can be 500 km.

When the wave approaches towards the shallow water, tsunami speed is noticed to get decrease and the energy gets transformed into a wave with higher heights. Sometimes, this wave reaches the height of 25m. But, the interval of time that takes place between the successive waves remains to be unchanged which is usually between 20 to 40 minutes.

When the tsunami reaches near the coastline, then the sea recedes to a level which is much lower than the low tide and then it rises as a giant wave.

The Pacific Tsunami Warning Center (PTWC) is considered to provide a warning for the pacific tele-tsunamis that means tsunamis that can cause damage far away from their source, to almost every country around the Pacific Rim and Specific Island.

Satellite or aerial photography is considered that they can provide critical information for the emergency managers when they are especially combined with the good GIS database of an area.

They even include damage to the structure, communication and transportation links, and including the life-line infrastructure components.

4.2.4. Hurricanes

Hurricanes are broadly known as tropical cyclones, which are rotating systems of clouds and thunderstorms. Hurricanes are considered to form over the tropical or subtropical waters.

One of the most powerful nature storms is hurricanes which can bring strong winds, heavy rainfall that can lead towards inland flooding, storms surge flooding, rip currents, and tornadoes.

These large scale low- pressure systems are considered to occur throughout the worldwide over zones and they are typically referred to as "tropical cyclone basins" (www.oas.org/usde/). The determination about the paths of the past hurricanes for the region is being recognized by deriving the data from the remotely sensed data collected by the U.S. National

Oceanographic and Atmospheric Administration (NOAA) satellite sensors.

The Tropical Analysis and Forecast Branch of the Tropical Prediction Center (TPC) help in providing year-round products that involve aviation forecast, surface analyses, marine forecasts, and warnings (SIGMETs).

Figure 4.3. Image taken from a satellite, representing the formation of Hurricanes.

Source: Image by Guard COAST GUARD COMPASS

The unit also helps in providing the satellite interpretation and satellite rainfall that estimates for the international community. The Technical support branch helped in providing the support for data processing of the satellite.

There was one of a key lesson that was learned by the NASA team during Hurricane Andrew was that it is found to be critical to select the appropriate data and put it together in order to make informed decisions. As this is a lengthy process that requires gathering the data, so it was suggested that communities do not wait until a disaster happens.

Imagery is considered to be an important aspect of the database community. With the help of the next generation of satellite- like Earth watch's Early Bird and Astrovision, there will be a significant enhancement in the remote sensing capabilities.

In the present time, the best sensor is the AVHRR that is used for plotting new data. AVHRR ranges about 2,490 km of swath; it has appropriate resolution and twice a day coverage. In this case, a red band is used for defining the daytime clouds and vegetation, whereas the TIR band is used for cloud observation, which is done both in daytime and nighttime.

4.2.5. Landslides

The movement of the debris, the mass of rocks or earth down the slope is defined as the landslide. Landslide is considered as a "mass wasting", which denotes any movement that takes place in soil and rocks in down-slope direction under the direct influence of earth gravity.

An area that already contains potential landslide hazard is usually considered to have some of the evidence that are related to previous occurrences. During the examination it be can be clearly seen by the stream traces which frequently show the deflection of the bed course which has occurred due to landslide.

It is believed that everyone is aware of the catastrophic landslides, but there are only a few who are aware of small slides that continue concern for those who are basically involved in the design and construction business.

Most of the landslide discussion does not focus on the problem of sinkholes, these are formed by the circular collapse landslides. The karstic areas in which it is seen to occur mostly are easy to identify using some satellite imagery like SPOT, MSS, TM, etc. This area is easily identified because of the pitted appearance and evidence related to essentially internal drainage.

There is a requirement for spatial resolution for recognizing most of the large landslides features which are about 10m (Richards, 1982).

However, it is believed that the recognition depends on the ability and experience of the interpreter and is also enhanced by the availability of the stereoscopic coverage.

Stereoscopic coverage and requirement for the resolution often preclude the use of most of the satellite-borne sensor imagery. The large block landslides can be easily detected by using Landsat, MSS and TM imagery.

SPOT HRV-P imagery which is in the panchromatic mode which is provided in the spatial resolution requirement is found to be useful with its 10m resolution.

As there is the presence of broad -band coverage, it is not conductive that can provide adequate contrast in the scenes which is involving heavily vegetative tropics, where most of the hazards occur.

Detection of landslide features is considered to be more easily achieved which is using airborne sensors. It is considered that Aerial photography with the normal stereoscopic coverage is one of the best sensor systems which can be used to define the landslides, both small and large.

Aerial photographic scales are considered to be as small as 1:60,000. Black and White panchromatic or IR films are considered as adequate in most of the cases. But it is also proved, that IR is better in some of the instances.

The IR sensitive emulsion is used for elimination much of the haze that is basically found in the humid tropics. In the aerial IR stereo-model, it is being noticed the open water or other moisture takes place in the back of the recent slump features that stand out as an anomaly, which can be either in black or in white color.

The color IR photography in some cases might be showing the stress that is seen on the vegetation which is caused by the recent movement. It is believed that if the scales are large enough then the tree deformation which is caused by the progressive tilting of the slope of soil can be also detected.

A thermal IR scanner is considered as the more sensitive detector of moisture which is associated with the landslides. The sensor is particularly useful in locating the seepage areas that are basically lubricating the slides. It is found effective particularly at the night when there is a maximum temperature difference that takes place between the effluent groundwater and the terrain.

There are many other factors despite its utility, which rule out the widespread use of thermal IR scanner. Therefore, these factors basically found to include the low altitude that is often required for reasonable spatial resolution. There are large numbers of flight lines that are required for the large area and for the geometrical distortions which are inherent in the system.

It is believed when the terrain is to be interpreted and it has some relief and is also non - descriptive, and then these distortions have become an even greater problem when the data is being interpreted by making the location of features difficult.

SLAR, is especially with the X-band synthetic aperture radar that is especially with its nominal 10m resolution, is found to be marginally useful in a stereo mode. This found to be in a stereo mode because of its ability to define some of the larger textures that are related to the landslide.

It is believed that in some of the cloud prone environment, radar may be considered as the only sensor that can be useful in providing interpretable information.

4.2.6. Floods

Flood is considered the most common natural hazards that can impact people, natural environment and infrastructure. It is believed that they can occur in many ways and in any environment.

The most prevalent is the riverine floods which occur due to heavy and prolonged rainfall, the appearance of rapid snowmelt in the upstream watersheds or the regular spring thaw.

Another flood is also caused by the extremely heavy rainfall, which is occurring over a short period in relatively flat terrain.

The backup of estuaries is seen to coincide with the storm surges due to high tides, dam overtopping which occur due to landslides into a reservoir. Even it is being noticed that the seiche and wind tides affecting the larger lakes.

Figure 4.4. An image representing the effect of flood.

Source: Image by Offutt Air Force Base

Occasionally there is an eruption on a glacier or snow-covered volcanic peak. This can cause a flood or a mudflow in which the terrain is found to be radically changed and if there is any agrarian development that is totally destroyed, frequently with loss of life in large quantity.

It is found to be impossible to define that there is entire flood potential that is present in a given area. However, the best remote sensing data that is suitable for this situation and a competent interpreter, the evidence that is used for potential flood situations that can be found or inferred.

Outside the historical evidence, the most related evidence of a major flood potential is the identification of those areas which are floodplain or flood -prone. These areas are generally recognizable by using remote sensing imagery. One of the most valuable applications of remote sensing related to flood hazards.

According to the Federal Emergency Management Agency (FEMA), floods are considered the second most common and widespread natural disaster. Within the USA, there are about more than 255 people who are killed and there is more than $3.5 billion of property which was damaged by heavy rainfall and flood.

There are many scientists and researchers who have been investing their valuable hours and funds in order to find a more accurate and faster method that can help in predicting and estimating the flood depth and their extent. Satellite imagery can be found very effective for flood management in the following ways:

- Development of a larger scale view of the general flood situation within a river catchment or coastal belt with the aim of identifying areas at greatest risk and in the need of immediate assistance;
- Monitoring of land use/cover changes over the years, in order to quantify prominent changes in land use/cover in extent and general of impervious area in particular; and
- Detailed mapping is required for the production of hazard assessment maps and for input to various types of hydrological models;

Generally, the flood- prone and flood plain areas can be identified simply by using remotely sensed imagery (Brakenridge et al, 1998). There are basically two types of efforts that have been made in these areas:

- flood mapping (Barber et al, 1996; Okamoto et al, 1998; Saatchi et al, 2000), which is using images of peak flood/post-flood (with water levels clearly visible) and

- flood forecasting, which is mainly based on cloud patterns (Pankiewicz, 1997; Feidas et al, 2000).

Flood Mapping: flood mapping is basically done by simply overlaying pre-flood images and a peak flood image. This is done in order to delineate the inundated area.

Whereas, when it comes in terms of properties and crops, damages are being calculated with the help of pre -existing land use base map (Colby et al, 1999; Brakenridge et al, 1998; Nico et al 2000; Saders and Tabuchi, 2000).

Nico et al, 2000 recently employed the amplitude change detection method that is used to detect the flood inundated areas by using multi -pass SAR data. Therefore, both the above -mentioned methods are used to identify the flooded zones.

These methods allow this identification only when they are flooding at the same time when there is a pass of a second satellite. But this identification is not possible for the first time.

Flood Prediction: flood prediction is done by remotely sensed data that is used in flood management. This flood management is possible when the flood is based on the presence of rain- bearing clouds. Flood prediction is totally based on monitoring of the cloud that is associated with the heavy rain and that can be achieved using satellite imagery.

This analysis is also supported by the humidity and temperature profiles that are achieved from the radiosondes which are present in the geographical area as well as by the synoptic maps of the area.

4.2.7. Desertification

Desertification is a process that occurs when an ecosystem experiences a decrease or loss of productivity. This process is believed to have a natural or anthropic component. This process may reinforce each other by creating a synergetic effect.

It is believed that the degree of desertification risk is found to be directly related to certain natural conditions. These conditions can be soil, natural vegetation, climate, hydrology and topography, as well as to the intensity and the type of anthropic activity happening in the area.

Desertification is considered among the most serious problems of the region. Therefore, this trend often indicates the increasing need to consider desertification processes in the integrated development planning studies.

It is considered that Remote sensing, either it is airborne or space-borne, it provides a valuable tool that can be used for evaluating the areas that are subjected to desertification. Film transparencies, digital data and photographs are being used for the purpose of locating, monitoring and assessing the deterioration fo a natural condition that is given in a particular area.

Information related to these conditions an is easily obtained from the direct measurements or inferred that is achieved from the indicators which are considered as a key that is used for recognition of a desertification process.

Figure 4.5. Image of the land after desertification.

Source: Image by Wikimedia Commons

For evaluating, describing and deciding about the type of action that is necessary to be taken, the following issues can be addressed by:

- Monitoring: this method is accomplished by detecting and measuring the changes which occur in the landscape characteristics, over a period of time. Comparisons are often made between previously observed conditions and present conditions for the purpose of recording the reduction in biological productivity.

- Assessment: it involves the identification and quantification of vegetative soils, landforms, cover types and land-use change patterns. The direction of change, rate of change, and vulnerability to change in desertification patterns can be studied through this assessment.
- Location: it involves the identification of areas, that are currently undergoing desertification and with that also identifying the areas that are expected to be exposed to the forces that can lead to deterioration.

In order to have a more detailed approach, there are four other sets of data that should be taken into a consideration for a desertification study of a given area: a set taken at the end of the last dry season, a set taken at the end of the humid season, and rest two seasons are taken five or ten years earlier (López Ocaña, 1989).

Data selection for a given area is directly related to the desired amount of detail, required degree of precision, size of the area and accuracy, and available in a time frame.

When there is the use of large-scale aerial photography, then it is believed that they provide a great amount of detail for this type of study. Systematic reconnaissance flights can be used for the purpose of environmental monitoring and for the resource assessment.

Radar sensors and infrared scanners are used for monitoring the soil moisture and other desertification indicators. It is considered that the acquisition of this type of data is found too costly and time -consuming. During the first stages of a detailed desertification study, the use of satellite imagery is recommended as it offers an overview of the entire region.

There are some factors like computer enhancement, classification, and false composition that can be useful in offering the required information. Optical enhancement can be also performed, but these are believed to lack the quantitative control that is available through an automated approach.

Even the statistical data is being obtained from a quantitative analysis by simply using the geographic information system that can be used to express it as a histogram, a table, in a new image, or in a graph. AVHRR imagery has been used for vegetation change studies and it can be even commercially available.

Ground resolutions of about 1 to 4 km are being represented by some of the limitations by making the large continental area studies. Whereas, other studies have used Nimbus data in order to delineate the moisture patterns and boundaries of vegetation.

In order to effectively locate and measure the dust plumes, there is a use of Geostationary Operational Environmental Satellite (GOES) data. Seasat SAP imagery has been used in order to have the delineation of large dune morphology.

Landsat, TM, SPOT and MSS data are considered to be very useful and cost -effective for the regional assessments. Landsat transparencies of bands 5 and 7 can be used for monitoring the superficial changes in the areas which are undergoing desertification. They are also used for mapping the present water bodies and former drainage systems.

It is being noticed that temporal tonal variation that is on the Landsat MSS is correlated with the variation on the field. The movements of sand-dune belts have been detected by using the Landsat with multitemporal approaches.

Albedo changes in arid terrain are being calculated by using Landsat digital data. This is considered as a phenomenon that basically tends to lower the productivity, in turn, it increases the rate of erosion, deposition of colic sedimentation, loss of vegetation density. It also tends to appear brighter on the image.

On the other side, this phenomenon also tends to increase productivity, with it also tends to darken the land. In this way, over a period of the brightness variation can be detected in an area. These data can be also calibrated with the ground data which is collected from the areas where there is an occurrence of the change.

Aerial and space remote sensing methods are providing a valuable tool that is used for desertification studies. Whereas, for any of the other natural hazard- related studies, it is important that they must combine with the ground collected data.

The use of remote sensing methods should minimize the ground data collection methods; therefore, this will save time and will give a quick result which is inexpensive per unit of data. The combination of remotely sensed and ground collected data can then be helpful for providing the basis for the easy assessment.

4.3. ROLE OF GIS IN DISASTER MANAGEMENT

Geo-informatics is considered to combine the satellite geodesy. Information technology and Geographical information system (GIS) is said to be comparatively new technologies which are capable to assist in all the phases of disaster and disaster management.

- Prediction,
- Assessment of damages,
- Post- disaster management, and
- Prevention, monitoring and warning system.

These technologies can be also used for computer modeling in order to study complex spatial patterns, causes and effects of relationships, and distribution of and the likelihood of integration of multiple data layers. It is believed that the deterministic model is capable of explaining the real cause of these kinds of hazards.

In spite of having a tremendous potential of Geo-informatics application in disaster management, yet it is noticed that technology is yet to be used optimally only by the decision- makers. Ironically, it is believed that still there is very limited used of geo-informatics as a tool for research subjects.

Training and capacity building of disaster management is considered effective in utilizing the Geo-informatics which is an important area that specifically needed focus, besides the huge investments that are being done for achieving the technological milestones.

4.3.1. Role of Geo-Informatics in Drought Mitigation

Remote sensing data help in providing the major inputs to all the three types of rainfall predictions, which are namely, short term predictions, medium-range predictions and long-term seasonal prediction.

It is believed that the Global and regional atmospheric parameters, ocean and land parameters such as wind, pressure, snow, temperature, El-Nino, etc are often required for the long-term prediction.

This prediction can be generated from the observations that are made by the geo-stationary and polar- orbiting weather satellites like NOAA and INSAT.

In case of medium- range weather prediction, the National Centre Medium Range Weather Forecasting (NCMRWF) who often uses the satellite-based sea surface temperature, snow -covered area and depth, roughness, soil moisture at surface level and vertical sounding, surface temperature, normalized difference vegetation index, pressure and temperature, vertical profile data in the T86/NMC model, altitude, and radiosonde data on water vapor.

In the short-range rainfall prediction, there is also INSAT-based visible and thermal data that is being used.

4.3.2. Role of Geo-Informatics in Cyclone Warning and Management

For about thirty years, Meteorologist has been using the satellite images to have easy monitoring of storms. One of the most important applications that are being used to determine the strength and intensity of a storm is the endeavor.

Figure 4.6. An image representing the formation of cyclone.

Source: Image by Wikipedia

It was noticed that in the late 1960s, the meteorologist has begun to observe that tropical cyclone in more frequent intervals. The infrared sensor aboard the polar -orbiting satellite which helps in providing the day and night observation. Whereas, geostationary satellite helps in providing continuous coverage during the daytime.

It is being observed that in India, there is a very efficient cyclone warning system that is comparable to the best known in the world. The basic approach was to predict the track and intensity of the cyclone by using the conventional as well as satellite and radar -based techniques.

4.3.3. Role of Geo-Informatics in Flood Management and Mitigation

For mapping and monitoring flood events in the near real- time and operational mode, optical and microwave data which is collected from the IRS, Radar sat series and Landsat ERS of the satellite have been used for this purpose.

Information related to the inundation and damage that is caused due to floods is being furnished to the concerned departments, which will allow them to enable them to organize all the necessary relief measures and also to make a reliable assessment related to flood damage.

It is observed that owing to a large swath and high receptivity, WiFS data from the IRS-1C and -1D can hold a great promise in the flood monitoring.

The information is being acquired during the flood, pre-flood and post-flood along with the information about the ground.

Flood damage assessment is being carried out by the process of integration which is done by topographical, flood plain and hydrological land use/cover information in a GIS environment.

In addition to this, the data related to space- borne multispectral have also been used for the purpose of studying the post- flood river configuration and about the existing flood control structure.

This space- borne multispectral data has also been used for identification of the bank erosion- prone areas and drainage congestion, with identifying the risk zone of the flood. In order to improve the flood forecast, the incorporation of remote sensing input is used.

These inputs are like satellite- derived rainfall estimates, soil information, detail about the current hydrological land cover/land used in the rainfall runoff model.

4.3.4. Role of Geo-Informatics in Earthquake Studies

> Earthquake basically occurs by the abrupt release of strain that has been developed in the earth's crust. It is being observed that most of the zone of maximum earthquakes intensity and frequency occurs at the boundaries.

These boundaries are between the moving plates that basically form the earth's crust. Some of the major earthquakes occur within the interior of the coastal plates. Such places are in China, Russia and some places of south-east United States.

There have been many types of research that have been carried out in order to predict the earthquakes by using conventional technologies. But the result related to dates is still inconclusive. In order to have knowledge about the historic earthquakes, the seismic risk analysis is being used.

With this, the presence of active faults is a method that is used for location and designing the dams, power plant and many other projects in the seismically active areas. In order to detect the active fault, there is a use of Landsat-TM and SPOT images and Radar interferograms.

Area rocked by the Lander earthquake is being studied by using the ERS-1 SAR interferometry, which is considered to get matched extremely well with the model of earth motion as well as by the local measurement. Side- scan sonar system can be also used for detecting the active faults that are found on the seafloor.

In recent years, space geodetic technique and high resolution aerial and satellite data are being used for the prediction of the earthquake. It is being believed that the space geodetic technique with the Global Positioning System (GPS) helps in providing an accuracy of a centimeter over 1000km.

Therefore, this helps in measuring the surface deformations and also for monitoring the accelerated crystal deformations which are prior to the earthquakes with the required accuracy.

Satellite imagery is being used in a delineating geotectonic structure. Satellite imagery is also used for clarifying the seismological conditions in the earthquake risk zone.

In order to reveal the active movement of tectonics along with the faults, there is the use of accurate mapping of geomorphologic features that are adjoining the lineaments.

There is a use of space techniques that helped in overcoming the limitations of the ground geodetic surveys or measurements.

This technique has become an essential tool that is being used for assessing the movement or displacements which takes place along the faults or plate boundaries to even millimeter level for accuracy.

By using Very Long Baseline Interferometry (VLBI) which make possible to record the accurate movement of the plate of the order of centimeter this is along the baseline of hundreds of kilometers.

Likewise, satellite- based Global Positioning System (GPS) for monitoring the changes over time, it has emerged as a powerful geodetic tool which is considered as a key tool that is used for understanding the long-term geodynamical phenomena.

GPS is considered to be useful particularly while measuring the more complex deformation patter which is across the plate boundaries where the large and regional scale strains are build up. There have been uses of differential GPS which are used for measuring the plate movements, slips along the faults and likewise many other factors to an accuracy of the sub-centimeter scale.

4.3.5. Geo-Informatics in Monitoring of Volcanic Eruption

It has been noticed many times that the precursors of the volcanic eruption are being used for observing the various areas of volcanic activity.

There are many changes like ground deformations, hot springs and crater lakes as well earth tremors, changes in the composition of the gases that are emitted from the volcanic vents, and changes in the temperatures of the fumaroles are proceeding of a volcanic eruption.

Figure 4.7. An image representing volcanic eruption.

Source: Image by Wikimedia Commons

For the purpose of volcanic hazard assessment, there is the use of thermal infrared remote sensing. Therefore, deficiencies in the coverage and equipment often suggested that the thermal infrared has not been evaluated for the purpose of the surveillance of volcanoes.

According to the national remote sensing agency, it has been demonstrated that the potential of multi -temporal Landsat-TM thermal bad data is used in the surveillance of active volcanoes over the Barren island volcano. This method was released during March 1991 to September 1991.

From the last three decades, it has been noticed that the aircraft and satellite- based thermal infrared (TIR) data is being used for the purpose of detecting and monitoring many of the activities of volcanoes around the world.

There are also some alternative tools that are being used for monitoring volcanoes; these are repetitive coverage, regional scale and low cost of thermal infrared images that are being gathered by using satellites. Although, there is a use of the spatial resolution of NOAA environment satellite which is too rough to record the details of surface thermal patterns.

Even, plumes of smoke and ash which are collected from the volcanoes are being used detecting the useful planning the rehabilitation of affected areas. Therefore, the study shows that there is an upward migration of the magma from the earth's crust which is just before the eruption takes place from the volcanic cone.

Such premonitory signs are considered to be easy and quick that is used for detecting the aids of differential SAR interferometry.

4.3.6. Role of Geo-Informatics in Landslide Studies

Aerial photographs and large scale satellite images are being used for locating the areas with the use of incidence of a landslide. Further, the use of higher spatial resolution and stereo imaging capability of IRC- 1C and -1D that enable further refining of the location and also monitoring of the landslides.

There is number of studies that have been carried out in India which use satellite data and aerial photography in order to develop an appropriate methodology for the purpose of classification of terrain and preparing of maps that are used for showing the landslide hazards in many areas of India.

These types of studies have been carried out in order to use mostly aerial photography as it has a high resolution which enables them to have a mapping with the intervals of better than 2m in height. The availability of 1m of resolution data is achieved from the IRS mission.

This may be helpful in generating the contour maps at 2m intervals and thereby making space remote sensing a highly cost- effective tool in the landslide zonation.

Figure 4.8. Image representing Landslide.

Source: Image by National Park Service

4.3.7. Geo-Informatics in Motoring of Pest Attacks

It is considered as one of the successful programs that are using space technology for the risk assessment from pest or diseases that are found in Desert. For this purpose, the Locust Satellite Application project of the UN/FAO is being used, especially for the International Desert Locust Commission.

Temporal and spatial distribution of the desert vegetation and rainfall is being derived from the NOAAAVHRR whose data is been used for the identification of the potential Locust breeding grounds. This process is being utilized in some of the South Asia countries in order to have an effective pest control measure.

Improvement of the desert locust forecasting system is being tried by the help of satellite data that is being achieved by the locust warning organizations. This warning is narrow down by the potential breeding areas in order to undertake the aerial spraying for the purpose of arresting further growth of locust.

4.3.8. Geo-Informatics in Forest Fire

There are several thousands of hectares of forests that are being burnt annually. The reason behind this is manmade forest fires that are causing extensive damage to the forest wealth. It is considered that the behavior of the forest fire depends on three parameters: weather, fuel and topography.

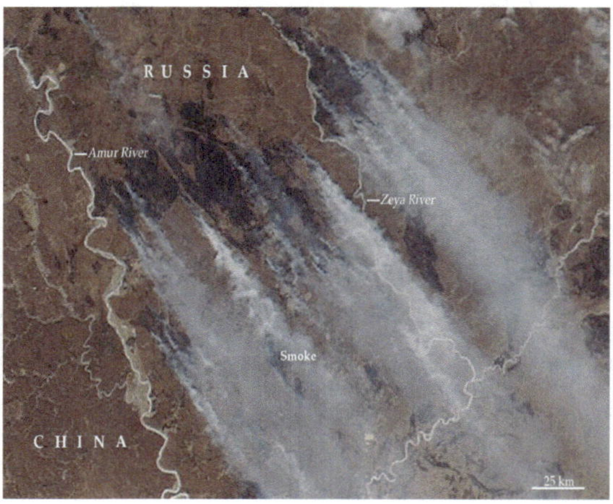

Figure 4.9. *Image showing a satellite view of forest fire in Russia.*

Source: Image by NASA Earth Observatory

Each of the parameters has several characteristic parameters. One of the most important tasks that are in the preparedness phase to assess the risk. In order to have risk assessment variables that are considered are such as land cover or land use, infrastructure, demography and urban interface.

There is a certain factor such as fuel mapping, rapid detection, assessment of the burnt area, identification of fire risk area and local and global fire monitoring are used as effective mitigation of forest fire.

The analysis of near real- time low spatial resolution of about 1km and high receptivity data that is received from NOAA and high spatial resolution data with the low receptivity from the earth resources satellite.

These all could provide information on the areas under fire. The IRS satellite data is being used for monitoring forest fires.

4.4. CONCLUSION

During the decade of the 1990's, this is being designated as the International Decade for Disaster Reduction. This has no resulted in the reduction of the losses which occur due to natural disasters. On the other side, the statistics show a rapid increase which is related to both, an increasing vulnerability of a large part of the earth's population and to an increase in the number of events related to weather.

This decade has also shown about the rapid increase in the capabilities of the technology and tools that can be used in disaster management. There are some of the tools that are used for dealing with the management and collection of the spatial data. These data can be related to Geographical Information Systems and Remote Sensing.

However, there is no new satellite that was specifically designed for using it for the purpose of disaster mitigation, whereas, there are most of them who have demonstrated their usefulness in the disaster prevention, relief and preparedness.

For several types of disaster, earth observation techniques are becoming operational in monitoring the phases and in the warning of phrases, which is within a short period of time. Remote sensing method is being used operationally for the purpose of warning and monitoring of the cyclones, drought and for the floods that have lesser extending.

The time that is required between the images that are being obtained until the answer is being given for specifying the warning or monitoring of the specific hazard, is called as the Turn - Around – Time (TAT).

TAT basically depends on many of the factors, like, the location of the event, weather condition (cloud- free images or radar), the satellite constellation, commercial and legal aspects, data analysis aspects like visual interpretation or automated analysis, and the data receiving aspects.

In many of the weather- related disasters, the process of obtaining the cloud- free images for the damage assessment which is often a severe problem. Some of the disaster types, such as oil spills, floods or debris flows, SAR is considered as a solution in that case.

Whereas, for another type of disasters like wildfires, landslides, earthquakes, optical images with the detail should be used. In terms of disaster relief, satellite remote sensing is believed to play an important role in the identification of the affected areas.

Figure 4.10. Image of satellite view.

Source: Image by Wikimedia Commons

It was noticed that with the use of the current system with the poor resolution, structural damage of the buildings cannot be observed. With the availability of the first civilian optical VHR (Very High Resolution) mission, near Real -Time Damage and the location of the possible victims is now possible.

Using IKONOS-2 will make the difference if the adequate temporal resolution, swath coverage and ready access data are being achieved. The temporal resolution which is being provided by the individual satellites is not sufficient and VHR will also not become operational in the mapping unless there is a use of multiple satellites.

Therefore, different GIS and remote sensing methods are being used for analyzing different disasters like drought management, cyclone, flood, earthquakes and many other natural disasters or man-made disasters. Nowadays, these are the most important techniques that are being used for analyzing the data that can be used for preparing the precautious measures for future prospective.

REFERENCES

1. 2014. *ROLE OF GEO-INFORMATICS IN DISASTERS*. [ebook] National Institute of Disaster Management. Available at: <https://nidm.gov.in/easindia2014/err/pdf/themes_issue/technology/role_of_geo.pdf> [Accessed 22 April 2020].
2. Krishnamoorthi, N., 2016. *Role Of Remote Sensing And GIS In Naturaldisaster Management Cycle*. 2nd ed. [ebook] Tamil Nadu. Available at: <http://www.onlinejournal.in/IJIRV2I3/025.pdf> [Accessed 22 April 2020].
3. Kulawardhana, R., 2012. Remote sensing and GIS technologies for monitoring and prediction of disasters. *International Journal of Digital Earth*, [online] 5(1), pp.88-90. Available at: <https://www.researchgate.net/publication/220473310_Remote_sensing_and_GIS_technologies_for_monitoring_and_prediction_of_disasters> [Accessed 22 April 2020].
4. Lwin, K., 2008. *Fundamental Of Remote Sensing And Its Application In GIS*. [ebook] Available at: <http://giswin.geo.tsukuba.ac.jp/sis/tutorial/koko/remotesensing/FundamentalRemoteSensing.pdf> [Accessed 22 April 2020].
5. Noaa.gov. n.d. *Hurricanes | National Oceanic And Atmospheric Administration*. [online] Available at: <https://www.noaa.gov/education/resource-collections/weather-atmosphere-education-resources/hurricanes> [Accessed 22 April 2020].
6. Oas.org. n.d. *CHAPTER 4 - REMOTE SENSING IN NATURAL HAZARD ASSESSMENTS*. [online] Available at: <https://www.oas.org/dsd/publications/Unit/oea66e/ch04.htm> [Accessed 22 April 2020].
7. Simonovic, S., 2002. *ROLE OF REMOTE SENSING IN DISASTER MANAGEMENT*. [ebook] London. Available at: <https://ir.lib.uwo.ca/cgi/viewcontent.cgi?article=1002&context=wrrr> [Accessed 22 April 2020].
8. Usgs.gov. n.d. *What Is A Landslide And What Causes One?*. [online] Available at: <https://www.usgs.gov/faqs/what-a-landslide-and-what-causes-one?qt-news_science_products=0#qt-news_science_products> [Accessed 22 April 2020].
9. Vijayaraghavan, C., Thirumalaivasan, D. and Venkatesan, D., 2012. *Utilization Of Remote Sensing And GIS In Managing Disasters*

– *A Review*. 3rd ed. [ebook] International Journal of Scientific &Engineering Research. Available at: <https://www.ijser.org/researchpaper/utilization-of-remote-sensing-and-gis-in-managing-disasters-a-review.pdf> [Accessed 22 April 2020].

10. WESTEN, C., 2000. *REMOTE SENSING FOR NATURAL DISASTER MANAGEMENT*. [ebook] Available at: <https://www.isprs.org/proceedings/XXXIII/congress/part7/1609_XXXIII-part7.pdf> [Accessed 22 April 2020].

Chapter 5

Role of Remote Sensing and GIS in Fishing Zone Forecast and Wildlife Habitats Evaluation

CONTENTS

The role of remote sensing and Geographical Information System (GIS) in the forecasting of Potential Fishing Zones (PFZ) is very useful for the fisherman. The chapter is all about how the use of remote sensing and GIS has helped the community in forecasting the PFZs. The first half of the chapter discusses the application of remote sensing in fisheries. The methods and materials, evaluation of remote sensing -based PFZs. The chapter in the mid explains the wildlife habitat evaluation, the role of remote sensing and GIS in the wildlife habitat evaluation. The chapter ends with the wildlife censusing and the role played by remote sensing in doing so. The utility of wildlife management is also explained at the end of the chapter.

5.1. INTRODUCTION

In order to ensure a profitable base of the economy, it is always economically significant that the forecasting of the Potential Fishing Zone (PFZ) to be carried for a sustainable environment. This helps in the sustenance of the existing fishing pool. The number of fish in the pool is affected adversely by the changes in the environmental conditions.

Earlier, boats and vessels were used to identify the potential fishing zones. This was the traditional sampling approach which was considered as costly and time- consuming. The reason being the vastness of the seas and oceans. The chapter introduces the effective methods of the forecasting of the fishing zones through airborne and satellite remote sensing.

The factors for identifying the potential fishing zones are the thermal circulation control in the sea and the concentration which is estimated from the estimated from ocean color dynamics i.e., chlorophyll. The Potential fishing zones advisories are prepared using the tools and the systems. These tools are also used to disseminate the information through the platforms based on Indian scenarios.

Due to the rich resource base, a wide range of biodiversity and the habitat which is productive, coastal environment plays a crucial role in the country's economy. There have been changes in the dynamic coastal fisheries sector over the years. It has also supported many people's lives (Nammalwar, Satheesh, & Ramesh, 2013).

The management and the development of the various seasonal stocks of the fishes in different areas are very important in the fishery sector. Through forecasting, the fishermen are guided towards those areas where there is large number of fishes found. They move to those areas to harvest most of the fish.

There are some commercial species that exist and they usually migrate due to changes in the environment. They move to favorable areas (Choudhury et al., 2007; Hela, 1960; Laevastu, 1984; Laevastu & Larkin, 1981; Laevastu & Rosa, 1963). As per FAO, 2009, most of the stocks of the fish are overexploited or fully exploited and the percentage is expected to be nearly 80%.

Therefore, the sustainable use of marine resources is essential for the effective monitoring and management of the fisheries. There are conventional potential fishing grounds that are located through:

- Fishing in traditionally known fishing areas
- A large number of sea birds found over an area
- Locating productive fishing grounds through breaks in color

The traditional sampling approach is surrounded by problems of cost and time. The use of boats and vessels in this approach is a time-consuming process and it involves huge costs.

The seas and oceans are vast and it is not possible to cover the ocean or sea for fishing purposes. However, it has been made possible through remote sensing. Remote sensing helps in monitoring the global oceans at equal intervals.

For example, the information related to the concentration of chlorophyll and the Sea Surface Temperature (SST) which is derived from the satellite images are widely used to describe the marine ecological provinces with physical, chemical and biological forcing.

The provinces are seasonally and an inter-annually variant with respect to time and space. The changes in the migrant sea fishes affect the distribution pattern, survival, condition etc. (Chassot et al., 2011; Klemas, 2013; Longhurst, 2010; Moore, Campbell, & Dowell, 2009; Oliver & Irwin, 2008; Stuart, Platt, & Sathyendranath, 2011; Stuart, Platt, Sathyendranath, & Pravin, 2011).

The spotter planes can easily and directly detect the group of fishes but it cannot be done through satellite altitudes. There are some biophysical indicators that can remotely detect and predict the potential fish aggregation areas. The images are taken from the satellite.

The indicators are the ocean fronts, the waters of the different ocean based on color, temperature. The upwelling areas that are more productive and greener than the background waters, certain fish species prefer zones of certain specific temperature (Klemas, 2013).

5.2. APPLICATION OF REMOTE SENSING IN FISHERIES

The role of the satellite is very important through which the advisories for the potential fishing zones can be found. This is the present scenario of the fishing sector. The use of the satellites has been increased for the purpose of fishing activities.

The fisheries resources could be studied with the application of remote sensing. By utilizing the sea surface temperature (SST) and chlorophyll parameters, the environment of the ocean, along with the availability of food in the marine ecosystem can be known.

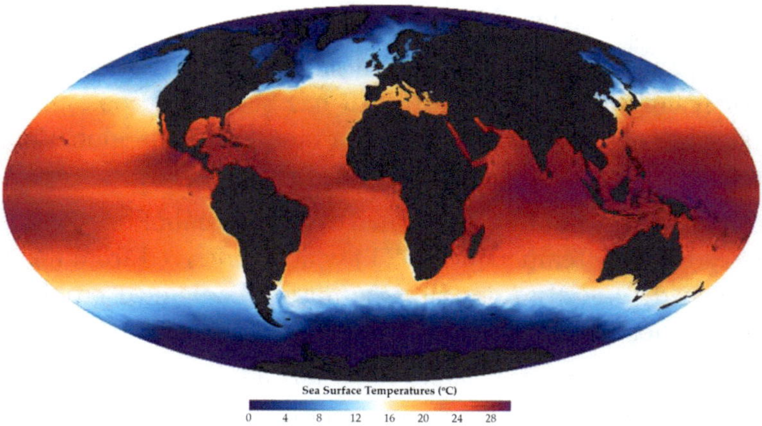

Figure 5.1. MODIS SST.

Source: Image by Wikimedia Commons

Or almost three decades, the SST images are being used for commercial purposes, especially for fishing operations. These SST images are satellite -derived (Breaker 1987).

The chlorophyll derived from Ocean Color Monitoring (OCM) and SST images derived from National Oceanic and Atmospheric Administration-Advanced Very High-Resolution Radiometer (NOAA-AVHRR) are being used for the exploitation of the fishery resources in the Arabian Sea (Solanki et al, 2003).

However, there are some disadvantages of using the images derived from SST and chlorophyll. The images from two different sources make it difficult to interpret the biophysical response of the oceanic surfaces. There is a time lag between the fishes being carried out from the sea and oceans in case of the Sea WiFS (satellite- borne sensor) and OCM.

In oceans, it is a well-known fact that the primary producer is phytoplankton and the secondary producer is zooplankton. Generally, it is believed that high productivity is the result of the high concentration of chlorophyll in the ocean (Sanjeev et al. 2011).

The fishery is influenced by the Plankton as they are responsible for the survival of the larval and also ensures recruitment (Pavan Dayaker, 2003). In the USA, the first applied fishery advisory through the satellite remote sensing was done in 1971 (Laurs, 1971).

The spawning in the waters of the California Bight of the northern anchovy was studied by Lasker et al. (1981), who used NOAA-AVHRR thermal and Coastal Zone Color Scanner (CZCS) ocean color imagery for the purpose of studying.

The technique of identifying the shoals of fish on the basis of certain oceanic features such as meandering patterns, eddies, rings, oceanic fronts and the upwelling areas is known as the Potential Fishing Zones. These upwelling areas are considered as prospective sites for the accumulation of fishes.

The Sea Surface Temperature (SST) generated NOAA- AVHRR Satellite data identifies these areas of fish accumulation (Solanki et al. 2003). The integration of ocean color with SST helps in the exploration of the fishery resources. This integration showed that these features are the sites of productive waters and therefore, it is conducive to fish aggregation (Dwivedi et al. 2005).

It is now very difficult to locate or catch a fish. This is emerging as a big challenge in front of the fisherman. The process is because time -consuming, costly and efforts are wasted.

The fishing community, therefore, needs a system or forecast of certain potential fishing zones or the areas of fish aggregation. This would reduce the time wasted in searching for the fish areas and their efforts would be saved in locating the fishing grounds.

As per the reports of International Union for Conservation of Nature (IUCN), 2010, there are 4161 of total terrestrial and aquatic species that are in threat due to climate change. There is a 33% risk due to habitat shifts, alteration due to climate change, 29 % risk due to temperature extremes and 28% due to drought.

The climate warming is indicated by the Cold-water fish salmonids and aquatic invertebrates. They are considered an excellent source of early warning indicators regarding climate warming. This is due to their body temperature.

The body temperature of the fish salmonids and the aquatic invertebrates depends on their surrounding and they have a low tolerance of thermal fluctuation.

Figure 5.2. Aquaculture of Salmonids.

Source: Image by Wikipedia

In order to compare the fish density/catch per unit of effort, the study was carried out. The satellite- based potential fishing zone product ESSO of INCOIS, Hyderabad (India), predicted the physical and biological parameters in the areas.

The electronic display boards were used to display the PFZ advisories to the fisherman at the Versova fish landing centers.

5.3. MATERIAL AND METHODS

There are some methods and material discussed below:

5.3.1. Generation of Fishing Forecast and Dissemination

The observations from the images were monitored and were placed in the composite images. The satellite images excluded features like eddies and meanders. However, the composite map made it possible to study the position of the PFZ with respect to its longitude and latitude.

The depth of this location can be read through images. The SMS and FAX facility was used for collecting the feedback data from the fishers. This information was disseminated to Versova through these facilities.

5.3.2. Experimental Fishing and Feedback Collection

A method which was used to collect the fishes was trawling net fishing for the North-West Coast. The selection of the trawl net fishing method was mainly done for the purpose of exploring the demersal region and for generating the data on species composition from PFZ such as mackerel, sciaenids, pomfrets, penaeid, tuna, coilia, and non-penaeid shrimps. This is done to compare it between PFZ and non-PFZ areas.

Figure 5.3. White Pomfret.

Source: Image by Wikimedia Commons

In order to locate the exact area of the fishing zone, Global Position System (GPS) was used. For getting the data on feedback, there were 3 – 5 trawlers that used to go for fishing and collect the data on the same.

The main purpose of feedback back data was to collect some information such as catch composition, the total weight of the trawl fishing in each species of fish, fishing depth, actual fishing position, number of nets used for fishing and hauling time.

5.3.3. Analysis of Water Quality Parameters and Catch per Unit Effort (CPUE)

The analysis of the water quality was done by collecting some nutrients from the water quality parameters. These parameters include total suspended material (TSM), pheophytin, dissolved oxygen, pH, temperature, salinity and chlorophyll.

The samples were collected from the PFZ area and later on analyzed in the research lab. Thereafter, these were plotted on a graph and the comparison was made between the PFZ and non-PFZ areas.

Generally, the potential fishing zones were located at the depth of more than 15–25 m below the sea and ocean. This is as per the PFZ advisories.

The distance of this from the shore was more than 25 km and sometimes extends up to 290 km. The PFZ data received from ENSO-INCOIS was closer to the shore which was about 25–30 km and depth of 15–25 m. Considering the availability of the vessel and its depth and distance, all the advisories of PFZ could not be studied.

5.3.4. Physicochemical Parameters of Water

The INCOIS PFZ advisories were regarding the collection of water samples from the specific latitude and longitude. For making the comparative study, water samples were also collected on the same day from the non-PFZ areas.

As and when the samples were collected, the parameters such as temperature, hydrogen ion concentration (pH), depth, D, salinity were measured onboard. There was a separate water collection for the purpose of estimating the nutrients, DO, total TSM, and chlorophyll.

The samples were collected in the respective bottles of the above. Moreover, the samples so collected were according to the guidelines for sample collection and preservation (Fodrie et al. 2010).

The water sample so collected was stored in the icebox at 4 °C for the purpose of transferring it to the laboratory for further analysis. These samples were collected in 500 ml polyethylene bottles. The standard procedures were followed to estimate the water quality parameters.

Digital thermometer and the mercury thermometer both were used for recording the temperature of the water. Measurement of the pH was done using the portable digital pH meter. Winkler's titrimetric method came to be useful for measuring the DO content of the water (Fodrie et al. 2010).

The vacuum filtration determined the TSM by using the pre-weighed glass fiber filter paper. The cadmium reduction method was used to analyze the Nitrate nitrogen (APHA, 2005). The guidelines of APHA, 2005 were followed for estimating the Chlorophyll and pheophytin and the results are expressed in μg/l or ppb.

5.4. APPLICATION OF REMOTE SENSING TO FISHERIES

As we know, Remote Sensing has been primarily used for assisting in the effective and efficient harvesting of natural resources but nowadays, its usage has been diversified and it is now used for better resource management, conservation and exploitation of resources.

Remote sensing can be of great use when it comes to Fisheries. The observations collected by remote sensing from the sea's surface can provide vital information with respect to improvements and assessments required in order to improve the potential yield of the fishing grounds.

As the sea covers two -thirds of the earth's surface, it serves the human population to a great extent and thus people are dependent on it for their food requirements such as aquatic plants and algae, fish, marine mammals, etc.

In order to exploit such resources more effectively, more fishing activities are required but at the minimum cost and better resources. This should be achieved keeping in mind all the restrictions specifically biological restrictions on catching fishes as some of the aquatic animals might be harmful too for the consumption of the human population.

Remote sensing can help to assess some particular knowledge required for the development of the Marine environment, for example, variations in productions and distribution level, currents, water circulations, regions of upwelling and concentrations of different matters whether dissolved or suspended. However, despite using different measurements by remote sensors, gathering all the information from the sea's surface is next to impossible which is needed to assess the marine environment as it changes rapidly and it becomes impossible to measure the entire range of information related to it.

Environmental conditions are constantly changing that not only affect the distribution system but also affects the availability of fishery resources.

The information collected can be used for forecasting marine habitats, however, the assessment of the presence of fish could not be collected through remote sensing, which serves as one of the disadvantages and improvement areas for remote sensors.

5.4.1. Direct Methods of Fish Detection

Some of the direct methods of Fish detection that are used widely are visual fish spotting, SONARs and Ariel photography. The simplest of them is spotting the different species of fish from the aircraft and then catching the most of them. This method of detecting fishes from the aircraft remotely is known as Visual fish spotting.

Another widely used direct fish detecting methods is the use of SONARs and echo- sounders that have been used from a very long time as a remote sensor, it emits sound waves underwater and when these waves get reflected from the objects underwater it produces sounds that help in measurement or presence of under-water marine animals.

Figure 5.4. Eco – sounding.

Source: Image by Wikimedia Commons

Aerial photography is another method wherein the pictures of areas with a major presence of aquatic animals or fishes are captured.

However, it is not suitable for the fishermen as it takes time to capture the pictures and then provide them to the fishermen as the marine environment is rapidly changing and it is not possible in a real time manner.

But Ariel photography is widely used among the scientists for researching the dynamic marine environment as it provides information about the distribution and relative abundance of different species of fish, particularly the schooling species. It also helps in estimating biomass of the species, distribution, location and the surface area of mobile fish spots.

5.4.2. Indirect Methods of Fishery Assessment

Indirect methods of fishery assessment use different combinations of various parameters to estimate the resources. These parameters might affect its distribution and availability, but it also helps the researchers to diminish the effects on environmental changes and it's after effect on marine animals.

Geographical and temporal distribution of fisheries resources is one of the major concerns of the researchers along with the constant environmental changes.

The fishes respond to a large number of environmental factors and changes their habitats constantly, thus, it becomes necessary to correlate various parameters to collect information about the fishery distribution through remote sensing.

The basic parameters include surface optical properties; diffuse attenuation coefficient yellow substance, chlorophyll pigments and macrophytes. These parameters are used individually or in different combinations by the researchers to assess the fishery information.

Surface Optical Properties

One of the standard methods of remote sensing for Fishery is said to be Surface optical properties. This is because of the presence of matter whether dissolved or suspended on the sea surface, which helps in determining the optical properties of the sea surface.

It is believed that under a normal environment, any visible light penetrates the water bodies' depth which makes the water more turbid. Now depending upon the properties of the substance concentrated in the water bodies, this light will be absorbed, scattered and reflected and thus can be useful in determining various information like imaging watercolors, sampling for further researches, and so on.

There are many parameters based on this theory of water emergent radiation using empirically constructed algorithms, some of them are discussed below.

- Diffuse Attenuation Coefficient

The diffuse attenuation coefficient refers to the measurements with which these radiations disappear in the depth of the ocean. It is specifically understood as an optical property. The speed of the same depends upon the distribution of radiations as a result of spreading, scattering and absorption at the point of measurement. This parameter in the combination of other parameters can be used physically to categorize the ocean by its color and its value can be determined based on the water turbidity.

- Total Suspended Matter (Seston)

Another parameter could be identified by the color of the sea surface. There are various agents that absorb lights and reflect lights affecting the color of the ocean. This parameter can be used while classifying organic and inorganic matter that settles at the bottom of the sea. This sediment concentration at the bottom of the sea could also be used for tracing water movements.

- Yellow Substance

Another parameter used in remote sensing is known as the Yellow substance which helps in the detection of polluted sea areas. The substance gathered on the sea surface due to degradation of land and marine organic matter is termed as Yellow substance. It helps in identifying the sea areas where the exploitation of fishery resources could be hazardous for human consumption.

- Chlorophyll Pigments

The photosynthetic pigments of phytoplankton are known as Chlorophyll pigments whose concentration in water bodies indirectly signifies the fish production. It is often considered as the barometer of biological productivity in the marine environment, usually referring to fish production.

Chlorophyll concentrations above 0.2 mg/cu.m indicate the presence of enough planktonic life to sustain a viable commercial fishery (Gower, 1972). These concentrations affect the ocean's color as they absorb red light and reflect the green light.

- Macrophytes

Macrophytes are commonly known as Seaweeds. These seaweeds are found in different regions in various types specifically in coastal areas, it is very easy to spot macrophytic vegetation. These seaweeds may reflect differently to remote sensors whether airborne or spaceborne sensors due to its various species it may reflect more green or red radiations, hence, it is suggested to use more airborne sensors like Ariel cameras to capture these reflections.

Figure 5.5. Aquatic Macrophytes.

Source: Image by Flickr

5.5. EVALUATION OF REMOTE-SENSING-BASED POTENTIAL FISHING ZONES (PFZS) FORECAST METHODOLOGY

The forecasts of PFZs that are Potential Fishing Zone which is satellite-based were generated by using the integration of OCM, which means Ocean Color Monitor, derived chlorophyll concentration and AVHRR that is Advanced

Very High -Resolution Radiometer derived sea surface temperatures which are denoted by SST.

During the seasons 1999–2000, 2000–2001 and 2001–2002, the validation of the PFZ forecasts was carried out by using the Fishery Survey of India's vessels. The bottom trawl catches contained mixed species from a number of various habitats, i.e. demersal and pelagic.

Therefore, it is important to understand the application of satellite-derived PFZs to various types of species of fishery resources. The contribution percentage of species was determined from 30 m to 100 m depth range. There were species that contributed 5 kg/haul and above and these species were said to be as a significant species.

Species-wise CPUE that is Catch Per Unit Effort, seasonal mean CPUE and SD i.e., Standard Deviation were computed for remotely sensed PFZs and the entire region. There were comparisons that were species-wise and these comparisons were made between mean CPUE in PFZs and seasonal mean CPUE in some other areas from a different habitat.

The observations were done and it was observed that the fishery resource of pelagic and water column habitats provided their contribution fairly well in comparison to the demersal resources in PFZs.

The contribution percentage of pelagic and water column species were found more in PFZs catch when it was compared with the seasonal mean catch in some other areas.

In this study, the evaluation of the diet composition and feeding habitat of fishes and its relationship with the satellite-derived parameters has been done.

5.6. WILDLIFE HABITAT EVALUATION

The term 'wildlife' applies to all the biotic elements which contain each and every species of plant as well as animal in this world. This does not include a man and his domesticated pets. In earlier times, this term was used most of the time for addressing a limited number of spectacular animals who tend to enjoy public recognition (Chakraborti, 1990).

In the world, the wilderness areas and wildlife habitats are shrinking in time and space due to unprecedented pressure of anthropocentric economic development.

Figure 5.6. Wildlife Habitat in Black Hawk County.

Source: Image by Flickr

In nature, wildlife and human beings coexisted until unless the humans became dominant organisms. Considering the modern civilization it can be said that by clearing the forests for settlements, agriculture and communication purposes and by setting up large hydroelectric projects, the industries have done much and irreparable damage to the natural system.

In many recent years, there has been much interest in the protection and conservation of wildlife which has been shown at different levels. There is a realization that is biodiversity holds the key for the wellbeing of humans and this realization has attracted the attention of the global and national organizations.

There have been debates going-on on the conservation and sustainable use of this wealth in the wilderness. Because of this, the planning activities are hampered because of non – availability of good quality data on wildlife, biodiversity and species habitats.

In the developing world where wildlife and biodiversity conservation is subordinated most of the time to more pressing needs and demands including hunger and poverty, the problem is more acute.

For inventory and evaluation of the natural resources, the quickest way possible is by applying remote sensing and GIS technologies.

The data provided through the remote sensing tends to provide the capabilities for synoptic, real – time, frequent and accurate assessment, management and monitoring of the natural resources. Management of natural resources and wildlife has become much more scientific and realistic along with the advancement taking place in the ecological and technological fields.

Now – a – days, the wildlife managers have access to a highly accurate spatial database. This allows them to consider the species – habitat relationship in a much better way.

There are a number of organizations such as IGBP that is the International Geosphere Biosophere Programme, WRI that is the World Resources Institute, FAO that is the Food and Agriculture Organization and WCMC that is World Conservation Monitoring Centre. These organizations are already using remote sensing regionally, locally or globally for a number of different purposes.

In the world, a number of countries tend to have their own remote sensing satellites which have up to 1m spatial resolution and facilities for data reception, data processing and data analysis.

An example can be of India who has already launched many remote sensing satellites and has also built many ground stations worldwide in order to receive the data (Anonymous, 2001).

5.7. REMOTE SENSING AND GIS IN WILDLIFE HABITAT EVALUATION

The management of the complete ecosystem is wildlife management (De Wulf et al. 1988). The sound wildlife management policies are the analysis and the quantification of the impacts on wildlife habitat such as agriculture, road development logging. This plays an important role in the phase of the process of making it a sound management policy.

There are many conventional techniques that have been applied for the collection of data on natural resources. There are large numbers of ground-based studies that are applied on the wild animals and other habitats and their movements in the corridor are studied (Bhat & Rawat, 1995; Hobaugh 1984; Johnsingh & Joshua, 1994; Mishra & Johnsingh, 1996; Rodgers 1990; Saxena, 1986; Tiwari, 1986).

However, there has been an increase in the attributes of the habitats, identification of the new sites for the protected areas and the current status of the corridors. This is made possible due to the increase in the use of remote sensing (Kamat, 1986; Panwar, 1986).

The all-time useful methods are the ground survey methods such as the collection of droppings, trapping, investigations of feeding sites, counting animals and ground mapping of habitats (Giles, 1978; Kotwal & Parihar, 1988; Lamprey, 1963).

Nowadays, remote sensing can replace or supplement the tedious work of ground survey methods in many cases.

Moreover, there are many limitations of the annually done ground survey as the whole ground survey cannot be done at a time. The information collected may also not be accurate which is possible through remote sensing and is aided by a limited survey of ground.

The spatial data and its comparison with the temporal variations are essential for wildlife management. This is only possible through remote sensing. Remote sensing allows the comparison of the spatial data with the temporal one. The managers of the wildlife are using the topographic maps to generate the maps of their interest (Leopold, 1933).

The remote sensing techniques are complex as it has technical impacts and applications. It has however revolutionized the process of data gathering and map- making. Remote sensing can be applied to wildlife census, evaluation and the wildlife habitat inventory. The land cover mapping is considered as same as the wildlife habitat mapping (Lindgren, 1985).

The vegetation composition or density and the landforms which are the biotic and abiotic components of the surface features can be mapped. The measurement of interspersion of habitat components, the distance to other critical habitat and the extent of habitat types is possible (Best, 1984).

There is the addition of a temporal dimension to habitat mapping and change detection due to the series of repetitive series of satellites NOAA (National Oceanic and Atmospheric Administration), IKONOS, SPOT (Satellite Pour l' Observation de la Terre) and IRS (Indian Remote Sensing Satellite). For over a 10-year period, Manitoba parklands, the digitized aerial photos were used by Adams and Gentle, 1978 for monitoring the changes in waterfowl habitat.

The correlation worked well in the variations in the transparency in the photographic density on true color with intra-habitat parameters. The habitat assessment of the elk, reindeer, *Rangifer tarandus* (George et al. 1977),

Cervus canadensis (Bright 1984), kangaroo, *Macropus giganteus* (Hill & Kelly 1987) has been done by the use of digital analysis of remotely sensed data.

The nesting sites of American kestrel (*Falco sparverius*) were predicted by Lyon, 1983 by using the nesting habitat model in combination with Landsat digital image classification. An effective tool for re-introducing the program of the white oryx, *Oryx leucoryx* was the interpretation through the Visual Landsat image (Harris, 1983).

The better understanding of the environmental patterns and processes is achieved from the data arrived from remote sensing. Wiersema (1983) studied the habitat of ibex in the French and Italian Alps by using Landsat data.

Different habitats for goral, *Nemorhaedus goral* in Rajaji National Park, sambar, Great Indian One-horned Rhino, *Rhinoceros unicornis* in Kaziranga National Park and *Cervus unicolor* in Kanha National Park are evaluated using the remote sensing and geospatial modeling.

Figure 5.7. *Naemorhedus goral.*

Source: Image by Wikimedia Commons

Roy et al., 1995, Porwal et al., 1996 and Kushwaha et al., 2000, 2001 have used remote sensing and geospatial modeling to evaluate the habitats for goral, Nemorhaedus goral in Rajaji National Park, sambar, Cervus unicolor in Kanha National Park and Great Indian One-horned Rhino, Rhinoceros unicornis in Kaziranga National Park respectively.

5.8. REMOTE SENSING IN WILDLIFE CENSUS

The wildlife censusing can be done by aerial visual observations, aerial photography and ground surveys. Earlier, the methods used to census the wildlife habitat were track counts, mark-recapture techniques, audible indices, hunter kill data, pellet group counts and actual field sightings.

Many wildlife populations are captured in the census through vertical aerial photography and are considered the best method of censusing the habitats of wildlife. The aerial photography is not useful in the case of birds as the aircraft disturb the bird's movement which renders it difficult to capture the air photos.

The mammals are being censused using vertical aerial photography which includes antelope, sea lions, moose, elephants, elk, whales, beavers, geese, flamingos, penguins, ducks, seals and oystercatchers.

It is obvious that aerial photography cannot be used in censusing all the animals of wildlife population. Only those can be captured that are open in the light hours and they can be counted (Leuthold, 1976; Lillesand & Kiefer, 1979). As per the indication of Burkhalter & Kientz (1984, wildlife censusing can be done with the help of thermal scanning.

The use of aerial photographs for censusing sandhill cranes, *Grus canadensis* have been discussed by Ericson et al. (1983). In order to determine the sex ratio of mallard (*Anas platyrhynchos*), Ferguson et al. (1981) have said that on the scale of 1:2000, aerial photography is cheaper than the observation in the ground method.

For the detection of deer, it has been reported by Wyatt et al. (1984), that an airborne multispectral linear array scanner that operates in the visible and near- infrared wavelengths is useful. The existence of mounds on the surface of the earth can detect the Fossorial mammals.

In the case of the pocket gopher, Thomomys sp., and prairie dog colonies, Cynomys sp., aerial photography is best- suited method for censusing (Best, 1984). Through the Landsat images, the detection of warrens of hairy-nosed wombats, *Lasiorhinus latifrons* was made possible (Loeffer & Margules,

1980). However, the use of satellite data for censusing the population of wildlife is limited by sensor resolution.

One more way to census the wildlife animals is increasingly spreading and that is radio- collaring. Radio Collaring is considered as the best method to locate the animals. It is basically a radio transmitter attached to the collar of the animal. It can easily track animals, it may be fitted to the collar or hand-held, fitted to the vehicle, an aircraft or a spacecraft depending on the design and capabilities of a particular system. Tracking with the help of aerial and space platforms is considered the best way to track than hand-held receivers (Leuthold & Sale, 1973).

5.9. THE UTILITY OF GIS IN WILDLIFE MANAGEMENT

The Geographic Information System is useful in mapping the wildlife movements. The Geographic Information System has made its contribution in the wildlife management, its distributions, movements of wild animals and other habitats, their uses, patterns and processes. GIS helps in mapping and analyzing the wildlife and can provide valuable information to help in the process of making of management strategies (Lawler & Edwards, 2002; Harvey & Hill, 2003; Fornes, 2004; Gibson et al. 2004; Greaves et al. 2006; Shanahan et al. 2007).

The utility of the ecological management and research has been increased due to the development in the field of remote sensing and Geographic Information System (GIS). Global positioning systems (GPS) are also at its growing stage which has contributed its utility in the mapping of the wildlife (Guisan & Zimmermann, 2000).

Mapping, monitoring, modeling and analyzing the behavior and patterns of the wildlife population has been benefitted with the use of Geographic Information System (GIS). There are instances of the studies where the use of GIS has made its contribution in the determination of distributions and extents of habitats selected by nesting animals (McLennan, 1998; Maktav et al. 2000; Sims & Smith 2003; Fornes 2004; Smith et al. 2004; Beggs 2005; Urios & Martínez-Abraín, 2006).

GIS is also useful in identifying the nesting habitats that are potentially available (Lawler & Edwards, 2002; Harvey & Hill, 2003; Gibson et al. 2004; Clement 2005; Mathieu et al. 2006).

5.10. CONCLUSION

Remote Sensing and GIS has enabled the identification of the potential fishing zones in a better way than before. Traditionally, people used to use the boats and vessels to find the aggregation of fishing areas. They use to move around on boats to find those areas which were time- consuming and tedious work. Remote sensing and GIS has made it easy and convenient to identify the potential fishing zones.

With the use of a Global Positioning System (GPS), the identification is now easy. It is done through the mapping of the PFZs. The forecasting information is disseminated to the fisherman, according to which they can approach the areas where there is a large number of fishes for fishing purposes. The process has now become less time consuming and is cost -effective.

REFERENCES

1. Advances in Geospatial Technologies, 2017. Remote Sensing Techniques and GIS Applications in Earth and Environmental Studies. [online] Available at: <https://www.igi-global.com/chapter/role-of-remote-sensing-in-potential-fishing-zone-forecast/172715> [Accessed 23 April 2020].
2. Clark, R., Mathieu, R. and Seddon, P., n.d. Geographic Information Systems In Wildlife Management. [ebook] Available at: <https://www.doc.govt.nz/globalassets/documents/science-and-technical/drds303entire.pdf> [Accessed 23 April 2020].
3. Kamei, G., Felix, J., Shenoy, L., Shukla, S. and Devi, H., 2013. Application of Remote Sensing in Fisheries: Role of Potential Fishing Zone Advisories. Geospatial Technologies and Climate Change, [online] pp.175-186. Available at: <https://www.researchgate.net/publication/284770829_Application_of_Remote_Sensing_in_Fisheries_Role_of_Potential_Fishing_Zone_Advisories> [Accessed 23 April 2020].
4. Kushwaha, S. and Roy, P., 2002. Geospatial Technology For Wildlife Habitat Evaluation. [ebook] Available at: <https://www.researchgate.net/publication/229036756_Geospatial_technology_for_wildlife_habitat_evaluation> [Accessed 23 April 2020].
5. P, N., Satheesh, S. and Ramesh, R., 2011. Application Of Remote Sensing In The Validations Of PFZ Along The Coast Of North Tamil Nadu, India. [ebook] Available at: <http://eprints.cmfri.org.in/9578/1/IJMS_42%283%29_283-292_Nammalwar.P.pdf> [Accessed 23 April 2020].
6. Solanki, H., Mankodi, P., Nayak, S. and Somvanshi, V., 2005. Evaluation of remote-sensing-based potential fishing zones (PFZs) forecast methodology. Continental Shelf Research, [online] 25(18), pp.2163-2173. Available at: <https://www.sciencedirect.com/science/article/abs/pii/S0278434305001688> [Accessed 23 April 2020].
7. Beggs, 2005
8. Best, 1984
9. Bhat & Rawat 1995
10. Chassot et al., 2011
11. Choudhury et al., 2007
12. De Wulf et al., 1988

13. Dwivedi et al. 2005
14. Fodrie et al., 2010
15. Fornes, 2004
16. Gibson et al., 2004
17. Giles 1978
18. Greaves et al., 2006
19. Guisan & Zimmermann 2000
20. Harris, 1983
21. Harvey & Hill, 2003
22. Hela, 1960
23. Hobaugh 1984
24. Johnsingh & Joshua 1994
25. Klemas, 2013
26. Kotwal & Parihar 1988
27. Kushwaha et al., 2001
28. Laevastu & Larkin, 1981
29. Laevastu & Rosa, 1963
30. Laevastu, 1984
31. Lamprey, 1963
32. Lasker et al., 1981
33. Laurs, 1971
34. Lawler & Edwards, 2002
35. Lindgren, 1985
36. Loeffer & Margules, 1980
37. Longhurst, 2010
38. Lyon, 1983
39. Maktav et al., 2000
40. Mathieu et al., 2006
41. McLennan, 1998
42. Mishra & Johnsingh 1996
43. Moore, Campbell, & Dowell, 2009
44. Nammalwar, Satheesh, & Ramesh, 2013
45. Oliver & Irwin, 2008

46. Pavan Dayaker, 2003
47. Porwal et al., 1996
48. Rodgers 1990
49. Roy et al., 1995
50. Sanjeev et al. 2011
51. Saxena 1986
52. Shanahan et al., 2007
53. Sims & Smith, 2003
54. Smith et al., 2004
55. Solanki et al. 2003
56. Stuart, Platt, Sathyendranath, & Pravin, 2011
57. Tiwari, 1986
58. Urios & Martínez-Abraín, 2006
59. Wiersema, 1983

Chapter 6

Techniques of Remote Sensing and GIS in Earth and Environmental Studies

CONTENTS

The chapter named 'Techniques of Remote Sensing and GIS in Earth and Environmental Studies' basically focuses on how the techniques were advanced and further used for studying various aspects of earth and environment. The chapter describes the evolution and advancement which was done in the remote sensing studies in order to study the environment. This chapter mentions a table which includes the indices which are used and their formula for the purpose of environmental management and also describes how it is used and how the management is done. The relation between remote sensing and earth studies is also described in the chapter. The readers will get insights about the use of GIS in mineral exploration.

6.1. INTRODUCTION

The remotely – sensed data is used in natural resources mapping and also, it is used as a source of input data for the environmental processes modeling. The use of remotely – sensed data has been considered as popular in recent years.

Radiometric and spectral resolutions have considered remote sensing as the best source of data for application and study over a large scale. This is done along with the availability of remotely sensed data from a number of different sensors of different platforms having a wide range of spatiotemporal.

There are some studies also in order to identify the changes. These studies have shown that the modifications that occur in between data related to vegetation properties are well identified when the enhancement of images is done by using vegetation indices prior to making the difference between the images. (Fung, 1990; Coppin and Bauer, 1996; Radeloff et al. 2000, Rogan et al. 2002)

These kinds of methods tend to allow the identification of subtle changes that are related to density, growth conditions and composition of vegetation (Horablaga et al. 2012). This further depends on the purpose and the spatial scale of the project (Rogan et al. 2003). There exists one more area where the image analysis is used a lot ad is considered to be an attempt in order to assess the plant health and also, for discovering their potential diseases.

The aim of some other authors is to identify the weeds in a cultivated area that considered the crop rows. There are some articles that are trying to find the way for the identification of those lands where the amount of pesticide used is quite high. This usage level reaches the level where it is detrimental to both crops as well as people (Maxwell, 2011).

This type also contains the articles which use the indices of different vegetation in order to determine the chlorophyll content of the crop. In such a way, it helps in discovering whether the crops are healthy or not (Haboudane et al. 2002, Daughtry et al. 2000).

There are some other studies also which aim to identify the degree of water evapotranspiration from the plants. There was one similar study that was done for the region of Morocco in order to better prepare the residents for irrigation (Simonneaux et al. 2007).

These were the elements that were studied by using remote sensing and also these are aimed in order to obtain a better production per hectare and healthier plants. Besides these, there are some studies also which have been done in order to estimate the production of a particular crop.

Those people who were dedicated to these types of studies have also tried to develop the indices so as to do the estimation about the productivity of crops without even considering their type (Zhao et al. 2013).

Some other studies were also there who tried to focus on determining the productivity of a single crop such as wheat (Kowalik et al. 2014), corn crops and soybeans (Johnson . 2014) and further, including the production of grapevine crops (Serrano et al. 2012, Govederica et al. 2015).

Some other studies are also there which are based on trying to predict the yield per hectare as well as they are also trying to determine the way in which plants will grow and the difference between the amount collected within the same parcel. Only by using satellite images or aerial photographs have a very high spatial resolution, this can be achieved (Yang et al. 2013).

There are problems which concern environmental protection and these problems are very vast. In legislation, the term 'environment' has never been identified explicitly. Even no one can find a definition of environmental protection. But now anyone can find terms which are related to several different aspects of environmental protection. This happened after the environmental law is made by a patchwork of laws.

Considering the field of economic sciences, the definition of environmental protection can be given as activities and furthermore, it allows for the inclusion of measures which lead to cost savings such as energy efficiency measures. The definition of environmental protection in the U.S. Environmental Protection Agency (EPA) strategy can be given as controlling pollution.

> So, it can be said that the environmental protection refers to any activity which is done in order to maintain or restore the quality of environmental media by further preventing the emission of pollutants or decreasing the presence of polluting substances in environmental media.

It can contain:

- changes in consumption patterns
- changes in the characteristics of goods and services
- changes in production techniques
- prevention of degradation of the landscape and ecosystems
- recycling
- treatment or disposal of residuals in separate environmental protection facilities

The introduction of the term 'geographic information system (GIS)' is done in the latter half of the 1960s and this was done by Roger Tomlinson. There are a number of reference literature on geographic information systems and these literature use a number of different definitions that are formulated at various approaches. Some of them include GIS is an information system, GIS is a toolbox, GIS is an approach to science, GIS is a business.

In broader terms, the GIS contains some methods, organization, financial resources, technical means which include hardware as well as software, spatial database, and the people who seem to be interested in its functioning. The geographic information systems are also termed as spatial information systems (SIS) and also as land information systems (LIS). This terminology is related to the area and also, with the scope of detail.

There is one more GIS specification which has geo-information systems. These are the systems that handle the processes which are related to geo-information, or the information that is acquired through the interpretation of geospatial data that pertains to spatial objects related to the Earth's surface. In addition to this, the geo-information is termed geographic information, spatial information, land information, geospatial information and environmental information.

Now in today's scenario, there are independent GIS systems that are built increasingly rarely. Most of the time, they are considered to be one of the components of different specialist information systems. And hence, the concept of GIS technology finds application more often. The concept of GIS technology is construed as a set of methods and techniques which serve to build geographic information systems.

These are said to be the methods as well as techniques which include inter alia: geodesy, remote sensing, cartography and photogrammetry and also, GPS which is Global Positioning Systems. The application of GIS technologies is widely been done at all the scientific fields and practical activities.

GIS can be used for the purpose of environmental management. Its use also includes a broad spectrum such as a simple formula and visualization of pollutant concentration in the environment, visualization of the natural data as maps of animate nature resources and also spatial distribution of the pollutants that can be even in the air.

In addition to this, GIS is also commonly used in order to plan and implemented environmental management processes such as water divide areas, area usage modeling hazard monitoring or forest protection against hazards.

Various types of data can be retrieved in order to monitor the environmental status and this takes advantage of the sensors and remote sensing measuring instruments. Furthermore, this permits the comprehensive examination of the conditions and changes which occur within the natural environment and its valorization, revitalization and protection.

As per the view represented in literature, the GIS methods are used and further help in evaluating the natural resources. It helps in analyzing the landfill and waste dumping areas which include their footprints; helps in defining the border runs such as forest division and forest road networks; helps in planning investment locations, delineates the protected areas and developing the noise maps.

The application of GIS is done in order to analyze and image the data of random nature like disease incidence, pest gradation and flood risks. Considering the protected areas and beyond, it can be said that GIS takes benefit of identification of the susceptibility of the natural elements to human impact and furthermore, managing the forest as well as water resources.

The GIS is deployed in order to monitor environmental conditions. These conditions include Lake Eutrophication, inter alia and invasive plants. GIS is deployed for spatial planning in order to reflect the locations which are related to environmental assessment requirements.

There are advanced spatial analyses that serve to carry out the comprehensive as well as multi-factor analyses of past and present events. These analyses help in forecasting and examining future conditions which are considered to be extremely important for estimating the natural hazards.

6.2. EVOLUTION AND ADVANCES IN REMOTE SENSING SATELLITES AND SENSORS FOR THE STUDY OF ENVIRONMENTS

In remote sensing, there are eight distinct eras. Among them, some are running parallel in the time periods, and there are some which are distinctly unique in content to technology, application in science, the concept of utilization of data and the data characteristics. These distinct eras are described below:

- **Airborne remote sensing era:** During the first and the Second World War, the evolution of airborne remote sensing era took place (Avery and Berlin, 1992, Colwell, 1983). This was the time when remote sensing was mainly used for purposes including surveying, mapping, reconnaissance, and military surveillance.
- **Rudimentary space-borne satellite remote sensing era:** At the end of the 1950s, the beginning of the space-borne remote sensing era took place along with the launch of "test of concept" rudimentary satellites (Devine, 1993, House et al., 1986). The rudimentary satellites include Explorer 1 by the United States and Sputnik 1 from Russia. Soon, in the late 1950s, this was followed by the first meteorological satellite known as Television and Infrared Observational Satellite-1 (TIROS-1) by the United States (House et al., 1986).

Figure 6.1. TIROS – America's first meteorological satellite.

Source: Image by Flickr

- **Spy satellite remote sensing era:** The time, at which the cold war was at its peak, the spy satellites including Corona (Dwayne et al., 1988) were widely used. The collection of data took place, almost exclusively, for military purposes. The data was not digital, but that data was produced as the hard copies. Nevertheless, during the above 3 eras, the spin-off of the remote sensing was developed for military purposes and these spilled over to mapping and after that slowly into environmental as well as natural resources applications.
- **Meteorological satellite sensor remote sensing era:** The geosynchronous Geostationary Operational Environmental Satellite (GOES), Advanced Very High Resolution Radiometer (AVHRR) and polar-orbiting National Oceanic and Atmospheric Administration (NOAA) (Kramer, 2002), these all consisted in the

early meteorological satellite sensors. This was an era in which the data started being available in digital format and furthermore, the data were analyzed by using an exclusive computer having a combination of both the hardware as well as software. This was also an era, in which the global coverage became realistic and the environmental applications became practical.

- **Landsat era:** The beginning of the Landsat era started with the launch of Landsat-1 which was then known as Earth Resources Technology Satellite, in the year 1972. The launch took place carrying multi -spectral scanner (MSS) sensor. This was followed by some other path-finding Landsat satellites 2 through 3. These satellites carried MSS and 4 and 5 which further carried Thematic Mapper (TM).

The Enhanced Thematic Mapper (ETM+) sensor was carried by Landsat 7. During the launch, the Landsat-6 failed. The Operational Land Imager (OLI) was carried by Landsat-8 and it is planned for launch in the year 2011. Also, the Landsat era has equally good sun-synchronous Land satellites which include Indian Remote Sensing Satellite (IRS) of India and Systeme Pour l'Observation de Ia Terre (SPOT) of France (Jensen, 2000).

These satellites have high resolution (nominal 2.5-80 meter) and these satellites also have global coverage potential. At such resolution, currently, only Landsat is collecting data with global wall to wall coverage. By far, this is the most significant era which kick -started the true wide environmental application of remote sensing data locally as well as globally.

- **Earth Observing System era:** The EOS that is Earth Observing System era (Stoney, 2005, Bailey et al., 2001, Jensen, 2000, Colwell, 1983) started along with the launch of Terra satellite in the year 1999. This has brought in the global coverage, high level of processing (such as geo-rectified, at-satellite reflectance), frequent repeat coverage, easy as well as mostly free access to data.

There were sensors carried by Terra\Aqua satellites. These sensors include the Moderate Resolution Imaging Spectroradiometer (MODIS) and Measurements of Pollution in the Troposphere (MOPITT). These sensors have daily re-visit and various processed data. The applications of sensor data have become widely spread and thus, the applications have multiplied.

There were institutions and individuals who never used remote sensing; now they have started to take an interest in remote sensing. In addition to all this, the availability of the processed data in context to products such as land use\land cover (LULC) and leaf area index (LAI) has become routine. Now, in the current scenario, MODIS itself has 40+ products.

The active space-borne remote sensing sensors tend to use radar technology. These sensors also became prominent around this time and also, during the Landsat era, these sensors include European Radar Satellite (ERS), Japanese Earth Resources Satellite (JERS), Radarsat and Advanced Land Observation Satellite (ALOS). The use of the Shuttle Radar Technology Mission (SRTM) was done in order to gather data for digital elevation.

- **New Millennium era:** There was an era named as new millennium era (Bailey et al., 2001) which refers to highly advanced "test-of concept" satellites that were sent into orbit around the same time as EOS era. But there is a difference in the concepts and ideas. Basically, these are satellites and sensors for the upcoming next generation. These include Earth Observing-1 which carried the first space-borne hyper-spectral data. The idea of Advanced Land Imager (ALI) as a cheaper as well as technologically better replacement for Landsat is also very interesting.
- **Private industry era:** The beginning of the private industry era took place at the end of the last millennium and at the beginning of this millennium (Stoney, 2005). This era contains a number of different innovations. First and foremost is the collection of data in very high resolution (that is <10 meter). This is typified by Quickbird satellites and IKONOS.

Secondly, it is said to be a revolutionary means of collecting data. This is typified by Rapideye satellite constellation which has 5 satellites. This satellite has almost daily coverage of any spot on earth at 6.5- meter resolution in 5 spectral bands and this includes a red-edge band.

The third point is the introduction of microsatellites. Some of these under disaster monitoring constellation (DMC), is designed and also, launched by surrey satellite technology Ltd. for countries like Turkey, Nigeria, China, USGS, UK, and others.

The fourth point is the innovation that is done by Google Earth while making rapid data access to VHRI for any part of the World. This is done by a technology named streaming technology which makes it easy for even a non-specialist to zoom as well as pan remote sensing data.

6.3. INDICES USED FOR ENVIRONMENTAL MANAGEMENT

Table 6.1. Indices used and their formulas

INDICES	**Formula**
Normalized difference vegetation index NDVI	NDVI=(NIR-R)/(NIR+R)
Normalized difference water index NDWI	NDWI=(NIR-G)/(NIR+G)
Normalized difference moisture index NDMI	NDMI=(NIR-IR)/(NIR+IR)
Normalized difference built-up index NDBI	NDBI=(IR- NIR)/(IR+ NIR)
(Modified) Soil –adjusted vegetation index SAVI MSAVI	$SAVI = \frac{NIR - RED}{(NIR + RED + L)} * (1 + L)$ where $L = 1 - \frac{2*s*(NIR-RED)*(NIR-s*RED)}{NIR+RED}$ where s is the slope of the soil line from a plot of red versus near infrared brightness values $MSAVI2 = \frac{(2 * NIR + 1 - \sqrt{(2 * NIR + 1)^2} - 8 * (NIR - RED)}{2}$

Source: Table from Researchgate

6.3.1. NDVI - Normalized Difference Vegetation Index

The values of the NDVI tend to vary on the basis of the absorption of radiation by the chlorophyll in the red spectral area and also, its reflectance in the near -infrared spectral area. These values exist between -1 and +1. These values express the consistency of green vegetation.

The values which are close to +1 (those in the light tones) represent the high consistency of the vegetation. And they are specific to dense hardwood forests. The values which are close to -1 (those in the dark tones) is considered to be the land lacking vegetation along with soil or rock. The value 0 (that is in mid -tones) is related to the land meadows. It is said to be useful while mapping areas having vegetation, vegetation typology, the use of land, the health of vegetation and much more.

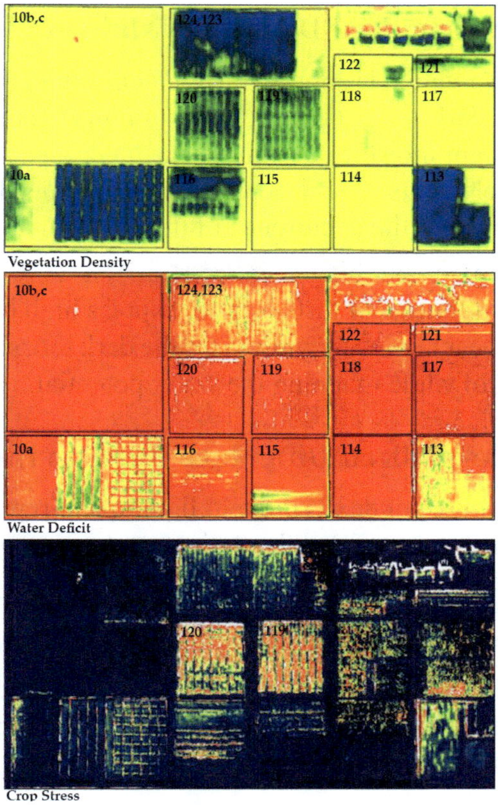

Figure 6.2. Image produced through Remote Sensing of Precision Farming.

Source: Image by Wikimedia Commons

6.3.2. NDWI - Normalized Difference Water Index

This index provides the turbidity differences and vegetation content of water, alluvial soils, or in the case when the differentiation of water content in vegetation is there. This index uses the green spectral bands (in which the electromagnetic radiation penetrates water) and near -infrared (which further increases the spectral response of moisture in soils, rocks and plants, and water starts to absorb radiation in the surface layers).

The values close to -1 are in the dark tones and these express the water surface. The values close to +1 are in the light tones and so represent the dry land. The values close to 0 are in mid- tones and it represents the lands having intermediate moisture content.

6.3.3. NDMI - Normalized Difference Moisture Index

In NDMI, the interpretation is done in the light tones (representing the excess moisture) and the dark tones (having low humidity). The NDMI tends to represent the different moisture content in the landscape and also, it shows ,in particular, the elements of soil, rocks and vegetation. This index is said to be a good indicator of the occurrence of the drought.

The values which are bigger in comparison to 0.1 and those are symbolized by light tones and furthermore, express high humidity. The low values (which are near about -1) symbolizes the dark tones, they express low humidity. It is useful while mapping the water potential.

6.3.4. NDBI - Normalized Difference Built-up Index

The interpretation in this index is done in light tones (representing the land with crops and constructions) and the dark tones (representing the land that is occupied by forests). The variation in the NDBI values depends on the spectral signatures of the middle infrared band (which show high reflectance of moisture in the soil, vegetation, rocks, including construction materials) and near- infrared band (which show high reflectance of chlorophyll). The positive values are represented by the lighter tones and these symbolize the land covered by buildings.

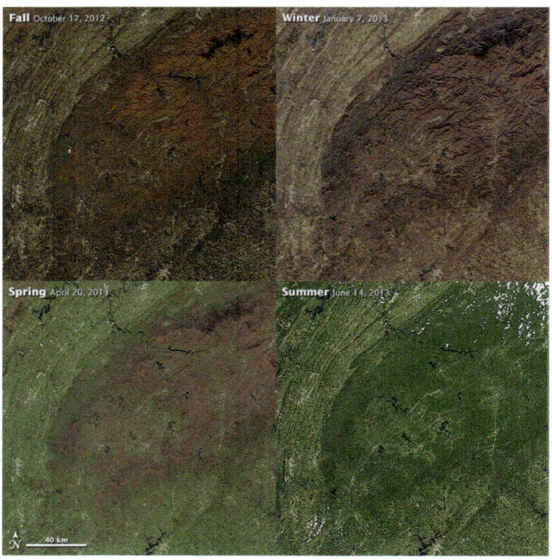

Figure 6.3. Satellite image in different seasons for a given area.

Source: Image by NASA Earth Observatory

The negative values are represented by the dark tones and these symbolize other landscape elements such as forests, farmland, etc. It is said to be useful in the automatic mapping of human settlements, and also, in some elements of the land such as railroads, roads, dams, canals, etc.

6.3.5. SAVI and MSAVI - (Modified) Soil –Adjusted Vegetation Index

In the current scenario, there is great interest in quantitative characterization of the spatial as well as temporal patterns of vegetation which further represent the data from a distance in order to study earth system science and global change.

The development of spectral models and indices is done in order to improve the sensitivity of vegetation while considering the effects of the atmosphere and soil. There are areas where vegetation topsoil is low, which means less than 40 percent, and the soil surface is exposed, in those areas the vegetation index values can be influenced by the reflection of light in red and near – infrared spectrum.

Particularly this is problematic when the comparisons are done on various soil types. Furthermore, this can reflect varying amounts of light in red wavelengths and near- infrared that include soils having different brightness values.

The adjusted vegetation index according to the soil was developed as a modification of the normalized difference vegetation index (NDVI) in order to correct the influence of soil brightness when the vegetation topsoil is low. The SAVI is structured in much similar to that of NDVI. There is a difference and that is the addition of a correction factor of soil light, MSAVI require one band red and one band near- infrared that is to be calculated.

The development of the adjusted vegetation index according to the soil (SAVI) was done in order to decrease the influences of the soil on the spectral canopy. This was done by incorporating an adjustment factor of the soil in the denominator of the equation that is for the NDVI which means normalized difference vegetation index.

The variation in the factor L should be done inversely in comparison to the amount of vegetation that is present. SAVI amended (MSAVI) for the optimum adjustment effect of the soil which in turn replaces the constant L in the equation of the SAVI to a variable function L. The derivation of the L function can be done by induction or by further using of the NDVI product and weighted difference (WDVI).

The aim of the MSAVI is to increase the dynamic range of the signal vegetation while decreasing the background influences of soil. This can further lead to the greater sensitivity of vegetation. There was a formula that was obtained and was known as MSAVI 2. This formula eliminates the need to find the ground line of a characteristic space or even it specifies a brightness correction factor of the soil.

The MSAVI calculation formula itself is the formula which is similar to SAVI. Generally, the MSAVI2 is considered to be the version that is used when people refer to MSAVI. There are several studies in which MSAVI was used where it was correlated most of the time with terrain data for vegetation cover, biomass and/or leaf area index, and also, as an entry layer for land mapping, classes cover or vegetation.

6.4. REMOTE SENSING AND EARTH STUDIES

There are two significant characteristics which the remotely sensed data have. First and foremost, they are available in digital form. This further allows them so as to be manipulated by the computer. The second point is that they tend to provide a range of information that is derived from the observations that are obtained by using EMR that is electromagnetic radiation. This EMR has wavelengths ranging from the visible wavebands (0.4 – 0.7 μm) to the microwave region (3-30 cm).

The modulation of the incoming EMR is done by the atmosphere or land or sea surface. This modulation provides information that is related to the characteristics of these land, ice and ocean surfaces and also, of the atmosphere. Let's take an example. When there are observations in the visible as well as near-infrared regions of the spectrum then they are used for the estimation of color (like those which are produced by the presence of phytoplankton in ocean surface waters).

There are thermal infrared as well as microwave measurements which can be used in order to monitor the surface temperature and roughness or soil moisture characteristics respectively. The overlapping or stereo images in the visible as well as microwave wavebands, are also used so as to derive the digital datasets further showing the surface elevation by using photogrammetric and interferometry techniques in a respective manner. There are certain products also which can be derived from these datasets such as maps of slope angle and aspect.

U.S. Apollo program was considered one of the greatest technical achievements of the twentieth century. A large number of people over the age of 40 years will recall the day on which Neil Armstrong was said to be the first human being to set his foot on the moon.

Now, in today's scenario, the impact of the Apollo program can be considered to be far wider in comparison to the purely technological, because of the fact that the photographic images of the earth taken by the Apollo astronauts generated an upsurge of interest in the human being and the society.

Geography has been considered as the study of "*Earth as the home of humankind,*" and the very first opportunity to view this home from outside was provided by the Apollo photographs.

The pictures or photographs which are provided by the Apollo astronauts helped in order to stimulate the growth of the "green" movement and also, the growth of environmental awareness among the public. The state of health of the oceans, the polar icecaps, the equatorial forests, and the atmosphere were then considered as political as well as scientific issues.

The public concern over "the environment" and this concern resulted in international agreements and also, in national legislation. Their intention was to control and also, reduce pollution of the atmosphere and oceans while decreasing the impact of global warming. The unmanned Earth-orbiting satellites such as the TIROS/NOAA provided the images and the weather satellites also gave their contribution to the public concern over the state of the environment.

It is evident that there have been considerable impacts of the images from the way in which human beings perceive their home planet. From a sixteenth-century sailor's point of view, probably the earth would have seemed very large as well as a hostile place which has unlimited resources.

When the human being entered the twenty-first century, their perception begins to changes and is quite different; they tend to see a small as well as fragile earth which has risk from the activities of its human population or inhabitants. This is a planet that requires care and attention. Also, human beings have learned that the atmosphere, oceans, land surfaces, and biosphere of the earth are all connected to each other intimately.

The Apollo photographs also had a scientific impact, along with the stimulating interest in the planet and the concern about planet Earth. There were pictures which are taken by astronauts at the time when the Apollo spacecraft were in low Earth orbit. These pictures demonstrated that valuable scientific information can be derived from the space images. The geological structures as well as drainage pattern, are much clearly visible for this remote region, which eventually covers an area that is far greater in comparison to what is shown on a conventional aerial photograph.

The ability to map the large areas of the surface of the earth was revealed by the images. Nevertheless, it was the potential ability of images from space platforms that they were able to track hurricanes and other weather systems. Ultimately this resulted in the first operational use of Earth observation by remote sensing.

In the year 1960, the first unmanned civilian satellite was launched by the U.S. National Oceanic and Atmospheric Administration (NOAA). This was specifically done in order to photograph the earth regularly which will provide the data for the purpose of weather forecasting.

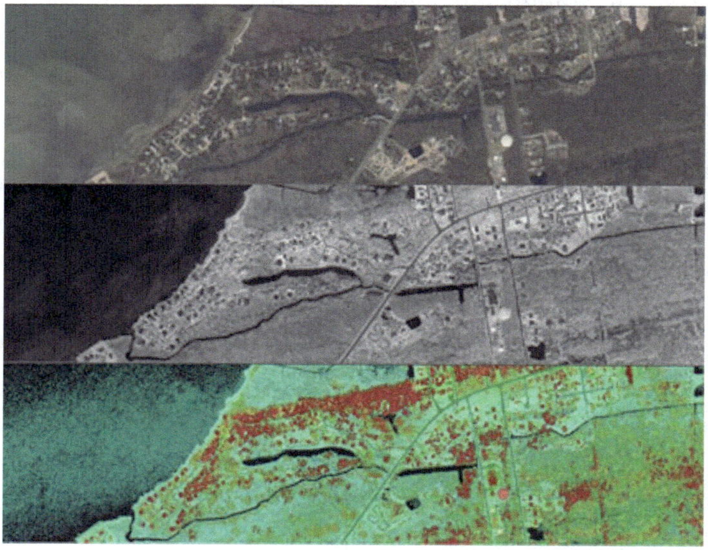

Figure 6.4. Visual Remote Sensing.

Source: Image by Flickr

This satellite named TIROS that is Television Infrared Observation Satellite was considered to be the first of the TIROS or NOAA weather satellites. These satellites are able to provide daily images of the globe, the images which are seen on the daily T.V. weather forecasts.

Both the polar-orbiting satellites including the NOAA and geostationary satellites like the GOES and METEOSAT are used routinely in today's era for the purpose of weather forecasting. The geostationary satellites tend to "hover" above a fixed point on the ground.

There is a wide area of view that can be obtained from the orbital altitudes. This is considered as one of the main reasons by which the images (that are obtained from instruments carried by satellite platforms) have been adopted readily while considering them as mapping and monitoring tools in disciplines including ecology, oceanography, geography, meteorology, and geology.

There is a second important feature of polar-orbiting satellites. This reason is that the configuration of them can be done in such a way in order to pass over the equator at the same local time each day, while further providing repetitive coverage of the surface of the earth.

There can be one example that can of the U.S. Landsat program. This program as initially designed as a monitoring system specifically for crops and vegetation. It is important to have frequent in order to observe the status of growing crops (the revisit time of the Landsat is 16 days).

While images from the space, there was one of the most important steps which was taken and that was the development of the instruments which further provided the data in digital form as well as in several bands of the electromagnetic spectrum.

It is important to have digital data if the computers are used for the purpose of the storage and manipulation of images that are collected by the sensors on the board orbiting platforms. This was done while the multi-channel data give more information regarding the nature of the targets which are observed on such images.

The electromagnetic radiation is used in order to sense the characteristics of the target, and also the interactions between the atmosphere and electromagnetic radiation, for a detailed description of the physical basis of remote sensing which includes the nature of remote sensing.

The remote sensing is only possible in some of the specific regions of the electromagnetic spectrum, in such kind of regions which are referred to as atmospheric windows. Considering the regions which are the visible and short-wave infrared (SWIR) regions of the electromagnetic spectrum, it is possible that the discrimination between the soil, water, and healthy vegetation can be done. Such kinds of graphs between reflectance and wavelength are known as the spectral reflectance curves.

There are some constraints that are imposed by the effects of atmosphere and the shapes of the spectral reflectance curves of targets of interest. These constraints mean that the instruments which are used for remote sensing for the monitoring of the Earth surface tend to use wavebands in the visible, SWIR, and thermal infrared wavebands (including the microwave bands). A large number of various measurements are done so as to discriminate between different spectral reflectance curves.

In the year 1972, the very first Landsat satellite was launched. This satellite carried a sensor that is referred to as MSS that is Multi - Spectral Scanning system. This is the type of instrument that contains four channels of data in which two located in the visible region of the spectrum (that is represented by the green and red) and two in the SWIR band.

It is possible to separate water from soil and water at wavelengths above 0.75 μm (micrometers). This is done so that either of the SWIR bands could be used for the same purpose. There is a spectral reflectance curve for vegetation which shows a characteristic peak at a wavelength of approximately 0.55 μm and an absorption feature at or near 0.7 μm, because of the chlorophyll absorption.

The green and red wavebands are used by the Landsat MSS sensor in order to pick up such type of features. There are instruments that are designed for other purposes, including CZCS that is the coastal zone color scanner, which tends to use wavebands that differ from waveband of the Landsat due to the fact that their targets are different.

The number and location of the wavebands are selected which are to be used in remote sensing. This selection is partly a function of the objectives of the program of work. Therefore, there is a system that is designed for oceanographic purposes and this system would acquire data in the wavebands which tend to differ from those which are used in the remote sensing of the terrestrial vegetation or in atmospheric monitoring.

In addition to this, it is also possible that such systems tend to differ in their spatial resolution. This can be regarded as the size of the smallest object for which a data value is being obtained. This smallest object in a digital image is referred to as a pixel.

Both the weather satellites as well as the ocean – monitoring satellites carry instruments that can view much of the surface of the earth in a single day because of the fact that the oceans and the atmosphere are dynamic. This means that their characteristics change rapidly over a period of time. In such sense, Amazonian forests are not considered to be very dynamic and thus, they are not a requirement of monitored each and every day.

The size of the image pixel is of the order of 1 km so as to collect the images of the surface of earth in a single day. If preparation has been done for accepting the lower temporal frequency of revisits for example, the 16 – day revisit period of Landsat in comparison to one day for the NOAA satellites, then it is possible to decrease the size of the image pixel.

The MSS of Landsat contains a pixel size of 80 m in comparison to 1 km for the AVHRR that is Advanced Very High- Resolution Radiometer that is carried by the NOAA satellite. In an image, the size of the pixel is connected with the spatial resolution of the sensor.

Only by using a coarse spatial resolution that can be of the order of 1 km, a daily global view is feasible. On the other hand, the accomplishment of the agricultural crop monitoring can be done by using a revisit period of around 14 days and also, a spatial resolution of the order of 30 m. it is important to have a higher spatial resolution for the purpose of mapping applications. The IKONOS satellite was launched in the year 1999. This satellite provides images having a spatial resolution as fine as 1 m.

During World War II, flight navigators reported that they gained the ability to see crude radar images of the ground below and to the side of the aircraft. From that time, the instruments have found wide applications in remote sensing. These instruments are those which have the capability of generating, transmitting and receiving microwave radiation.

There are certain multi-spectral and hyper spectral instruments which record the ground – leaving radiance in the visible as well as infrared spectral bands. Such kinds of instruments are referred to as passive sensors. The reason behind this is that they tend to rely on the sun or the earth for the energy that they record. Radar is said to be an active sensor and furthermore, can operate during both the time day as well as night.

Moreover, energy is able to penetrate clouds at the centimeter wavelengths which are used by the radar. Therefore, it helps in overcoming one of the major disadvantages of the optical radiation (visible as well as infrared). Seasat was launched in the year 1978 and is considered as the first radar satellite. During the 100 days of its operation, it obtained images of both the land as well as the sea which further proved to be of great value for the scientists.

From that time, there have been a number of space radar systems that have been placed in the orbit including the European ERS-1 and 2, the Canadian Radarsat and the Japanese JERS – 1. All of these systems tend to use a synthetic instead of a physical antenna and therefore are referred to as SAR that is Synthetic aperture radar systems.

They tend to use a single antenna and a fixed wavelength for their operation. There are experimental radar systems that have been carried by the space shuttle of the NASA in order to investigate the use of multiple wavelengths at the same time and of different directions of polarization. Furthermore, these experiments produced some results which will be implemented in the instrument systems that si to be flown in the early years of the 21st century.

There are a number of applications of SAR data and one of them is the development of measuring the elevation of the surface of the earth from space by using the technique of interferometry. It is important to have two SAR images of the same area of the surface of the earth which must be taken from different but close positions.

Then the surface elevation can be computed form the difference in the phase of the two signals and the distance that is apart in orbit of the two sensing systems. SAR data is not considered as the only source of remotely sensed data that is available for the purpose of mapping the surface models, despite the fact that it appears to be a cost -effective method for performing the detailed surveys in its early stages.

LIDAR that is Light Detection and Ranging data has one characteristic that are a profile of measurements can be obtained from the point where the laser light first reaches an object on the surface of the earth to the point where it reaches the ground.

6.5. PROCESSING AND ANALYZING DATA

It is important to have appropriate as well as a sufficient amount of data in order to make correct and accurate decisions. Despite the fact that there are some data which are straightforward and easy to understand, some other data including those with large quantity and sophisticates relationships to other data could appear to be much more intimidating, complicated and confusing. It is important to process and analyze these data so as to reveal the patterns and relationships hidden in the large amounts of numbers.

Let's take an example while reviewing the historical records and reports of the investigations done previously at an environmentally impacted sire, a large number of datasets much be gathered by an environmental scientist including some mentioned below:

- Cultural sites, habitats of protected species;
- Chemical results of soil, surface water, sediment and groundwater samples;
- Historical drawings and aerial photos;
- Current and planned future land uses, hydrology, geology, soil, vegetation and elevation data;
- Geophysical transects and anomalies and intrusive investigation results;
- Census data;
- Wetlands;
- Water wells;
- Right of entry (roe) information; and
- Affected properties, etc.

Nevertheless, it could be hard to locate the environmentally impacted areas, visualize their patterns and relationships, understand their changes that have been occurring over the time and delineate their extents on the surface and underground, by only considering these datasets.

It can be further difficult to even evaluate their current as well as risks that can possibly occur in the future to human health, the environment ecosystems and also, to design the remedial investigation strategies and technical approaches in a detailed manner.

The GIS that is Geographic Information System tends to offer the powerful data processing, data analyzing and modeling tools which further help the scientists and decision-makersers in visualizing and understanding the data in a much better way, in getting useful as well as more meaningful information from them and making some sound decision while solving the problems.

GIS also helps in delineating their boundaries as well as designing the investigative and remedial approaches which include obtaining ROE to the affected properties; deciding sampling types and locations; planning brush clearance; designing intrusive investigation grids and/or trenches; proposing geophysical survey methods, areas, and transects; and installing monitoring wells.

Figure 6.5. K-Box Support for GIS data.

Source: Image by Wikimedia Commons

In a similar manner, a historical drawing tends to show the detailed layout of installation and so, the historical drawing can be scanned and geo – referenced in order to identify the potentially polluted areas and some possible sources. These sources can include ground and underground fuel or munitions storage facilities; hazardous materials loading; unloading and processing areas; fuel stations; lands or dump areas; burial pits and maintenance shops.

Statistical analysis can be done of the analytical results of the samples and further, they can be displayed as color – shaded contour maps. These maps show concentration patterns throughput the whole project site and the changes that are occurring in the trends over time. Three – dimensional modeling and visualization techniques can be used in order to illustrate the data obtained while delineating underground soil and water contaminations.

The evaluation of the risks to the human health, the environment and the ecosystem can be done by analyzing the relationships possessed by the related data layers including samples, water wells, census data, wetlands, groundwater aquifers, rivers, lakes, streams, parcels, oceans, geology, habitats, land covers and uses, geology and topography.

Identification of the data clusters and data gaps can be done by statistically analyzing the existing data coverage of a project site and further the prediction of the situations in adjacent areas can be done. It is important to have additional data collection efforts for the data gaps in order to achieve sufficient data coverage of the whole project site while making accurate decisions.

Monetary savings can be achieved in the areas having data clusters and this can be done by avoiding the new data collecting efforts there. This can be further done by decreasing some existing data collection activities. These activities include the decommissioning the unnecessary monitoring wells and removing the unwanted sampling locations.

6.6. GIS IN MINERAL EXPLORATION

GIS is considered to be a digital database which further facilitates the integration of a number of different datasets for the purpose of spatial analysis and modeling by using a common spatial coordinate system. It can be said that GIS is majorly useful in handling the images, data tables, maps, analysis, visualization, modeling and spatial decision support. The identification of the mineral potential zones can be done by using GIS.

GIS is a tool that can be used in order to analyze the geological, geochemical, remote sensing data and much more. Mainly, GIS is used in the generation of mineral potential maps. Recently, there have a few authors who have discussed the role of GIS in mineral exploration (Bonham-Carter, 1994).

GIS that is a Geographical Information System is considered as a system of hardware, software and operational procedures which are very much useful in storing a large amount of information or database while editing and manipulating data and also interpretation and analysis of the larger volume of data. This data is provided as an input by the top sheets and the images provided by satellite of that specific area or region.

In addition to this, GIS can be used in facilitating while organizing several variable datasets for the purpose of querying, analysis. Therefore, it is helpful in drawing meaningful inferences from the data that is being generated. Also, it is useful while identifying a number of different parameters which further help in decision – making.

GIS also emphasizes the spatial context by further focusing on the geological features which can be localized mineral deposition. There are a number of authors who have studied the importance of GIS in mineral exploration (Wilkinson, et al, 2001).

Mine It is important to have data on the generation of thematic maps from various disciplines. This data needs to be integrated for the purpose of the zonation, integration and modeling of geoscientific data that is said to be useful while preparing the thematic layers. The technology is used for the ground -based data acquisition, information extraction and analytical GIS.

It provides its contribution in the rapid as well as efficient mapping of geology, structure and geomorphology in the area having limited IT infrastructure and also, in data by using some of the modern methods and also, by using geo-information technology (Chung, C. F., Jefferson, C. W. and Singer, D. A., 1992).

The Geographical Information Software is considered to be very useful while providing the input that is required and which is related to the geological resources of an area or region that is to be studied. It can be said that an important role is being played by remote sensing, and GIS while studying the mineralized areas by performing the evaluation of different thematic layers (including geology, structure, and geomorphology) by using spectral anomalies.

6.6.1. Role of Remote Sensing and Geographic Information System in Sustainable Development

It is important to have information related to nature, extent, spatial distribution and potential and limitations of the natural resources before planning the strategy for sustainable development. Along with this, it is also important to have socio – economic and meteorological and some other related ancillary information while recommending the locale-specific prescriptions in order to take up the curative or preventive measures.

Space – borne multispectral data have been used at the operational level in order to generate the baseline information on mineral resources, groundwater and surface water, soils, land use or land cover and forests, etc. this is done as a result of providing the synoptic view of a large area at a regular interval. This is done at scales that range from regional to micro-level that is 1:250,000 to 1:12,500 scale and also to monitor the changes which may have occurred over the period of time.

There have been indigenous effort which was put on the design and development of the satellites and sensors. These efforts have resulted in the launch of the Indian Remote Sensing Satellite that is IRS – 1A and B which carried the Linear Imaging Self – scanning Sensors that are LISS – I and II. These sensors have spatial resolution in comparison to those of Landsat MSS and TM, in the late 80s and early 90s.

In sensor technology, there were further developments taking place which resulted in the launch of the state-of-the-art satellite that is IRS-IC in the year 1995 which had a unique combination of the following three sensors:

- WiFS that is Wide Field Sensor with 188m spatial, two spectral bands - red and near- infrared, 810 km swath and receptivity of 5 days;
- LISS-III that is Linear Imaging Self-scanning Sensor having 23.5m spatial resolution in the green red and near- infrared region, and also, 70.5 m in the middle infrared region, and 140 km swath;
- PAN that is a panchromatic camera having 5.8m spatial resolution, 70 km swath and stereo capability.

Furthermore, there are certain advanced image fusion techniques which include HIS that is Intensity, Hue and Saturation transformation. By using these techniques further refinement in the information related to the natural resources can be done.

In a similar manner, when there are changes that have taken place either because of the developmental program or land degradation, then image differencing and principal component analysis is used for the monitoring of changes so as to provide more objective assessment of such type of changes.

Only optical sensor data having a few broad spectral bands have been used in order to generate baseline information related to the natural resources. There is hyper spectral remote sensing which has the ability to provide the diagnostic capability of some natural features which include vegetation, minerals, etc. This sensing will help in refining the information related to natural resources which have been generated.

The imaging of the terrain in the presence of smoke, haze and cloud cover is considered to be the major limitation of the optical sensor data. The microwave data having the day and night observation ability and also, the ability to penetrate through the cloud, haze, or smoke hole very good reputation while generating the information related to crop coverage, floods and much more.

The microwave energy having different polarization is used in order to generate the polar metric images and this further provides insight into the structure and floristic of vegetation, parent material and soil properties (Skidmore et al., 1997).

Furthermore, radar interferometry is considered to be another tool that enables generating DEM. This tends to allow the monitoring of volcanic eruption, glaciers, mine subsidence, mud slips, etc. it is important to do the integration of information related to natural resources, socio-economic and climatic conditions and some other related ancillary information in a holistic manner in order to prescribe the locale-specific intervention for a given area.

GIS that is Geographic Information System has the capability of integrating such spatial as well as attribute data and subsequent generation of action plan or developmental plan for the purpose of ultimately sustainable development.

The GIS has been used in a number of different applications, which include:

- assessment of surface and zonal models of the population (Martin,1996),
- database development and changes in the aquatic environment (Remilard and Welch, 1993),
- modeling of non- point source pollution (Welch et al, 1993),
- military housing management (Forgionne et al. 1996),
- database design for a multi-scale spatial information system (Jones et al.1996),
- Multiple criteria group decision making (Malczewski, 1996), etc.

There have been a number of studies that have used remote sensing and GIS conjunctively so as to address the issues which are related to developmental planning (Trotter, 1991; Smith and Blackwell, 1980; Welsh et al., 1992; Hellden et al., 1982).

6.7. CONCLUSION

From the above chapter, it can be concluded that a number of different computer hardware and software along with the latest technologies related to remote sensing and information technology is said to be much useful and also, it is referred to as highly important.

These technologies of remote sensing and GIS are useful in order to understand and also assess the present or current status of the existing geological resources in any provided area or region of the world.

The computer models have been used in order to interpret the past and present data and also, while in predicting the future status of such resources. Adoption and implementation of the techniques and scientific methods are being done in order to manage and conserve the geological resources in any given area of the region.

REFERENCES

1. Gajos, M. and Sierka, E., 2012. *GIS Technology In Environmental Protection: Research Directions Based On Literature Review.* [ebook] p.1. Available at: <http://www.pjoes.com/GIS-Technology-in-Environmental-Protection-r-nResearch-Directions-Based-on-Literature,88747,0,2.html> [Accessed 20 April 2020].
2. HERBEI, M., HERBEI, R., SMULEAC, L. and SALAGEAN, T., 2016. *Using Remote Sensing Techniques In Environmental Management*. [ebook] p.3. Available at: <https://www.researchgate.net/publication/311779226_Using_Remote_Sensing_Techniques_in_Environmental_Management> [Accessed 20 April 2020].
3. Mather, P., 2020. *REMOTE SENSING AND ENVIRONMENTAL MONITORING*. [ebook] p.4. Available at: <https://www.eolss.net/Sample-Chapters/C01/E6-64-02-00.pdf> [Accessed 20 April 2020].
4. Melesse, A., Weng, Q., Thenkabail, P. and Senay, G., 2007. Remote Sensing Sensors and Applications in Environmental Resources Mapping and Modelling. *Sensors*, [online] 7(12), pp.3209-3241. Available at: <https://www.ncbi.nlm.nih.gov/pmc/articles/PMC3841891/> [Accessed 20 April 2020].
5. Rao, D., 2000. *ROLE OF REMOTE SENSING AND GEOGRAPHIC INFORMATION SYSTEM IN SUSTAINABLE DEVELOPMENT.* [ebook] International Archives of Photogrammetry and Remote Sensing, p.2. Available at: <https://www.isprs.org/proceedings/XXXIII/congress/part7/1231_XXXIII-part7.pdf> [Accessed 20 April 2020].
6. Sangam, D., 2017. *RS AND GIS STUDIES IN MINERAL EXPLORATION - A Scientific Approach*. [ebook] p.2. Available at: <https://vignan.ac.in/vfstrstem/vol3/Narayan%20Sangam.pdf> [Accessed 20 April 2020].
7. Tian, B., 2016. *GIS Technology Applications In Environmental And Earth Sciences*. [ebook] p.2. Available at: <https://www.routledgehandbooks.com/pdf/doi/10.1201/9781315366975-4> [Accessed 20 April 2020].
8. Avery and Berlin, 1992
9. Bailey et al., 2001
10. Bonham-Carter, 1994
11. Chung, C. F., Jefferson, C. W. and Singer, D. A., 1992
12. Colwell, 1983

13. Coppin and Bauer 1996
14. Daughtry et al. 2000
15. Devine, 1993
16. Dwayne et al., 1988
17. Forgionne et al. 1996
18. Fung 1990
19. Govederica et al. 2015
20. Haboudane et al. 2002
21. Hellden et al., 1982
22. Horablaga et al. 2012
23. House et al., 1986
24. Jensen, 2000
25. Johnson, 2014
26. Kowalik et al. 2014
27. Kramer, 2002
28. Malczewski, 1996
29. Martin, 1996
30. Maxwell, 2011
31. Radeloff et al. 2000
32. Remilard and Welch, 1993
33. Rogan et al. 200
34. Serrano et al. 2012
35. Simonneaux et al. 2007
36. Skidmore et al.,1997
37. Smith and Blackwell, 1980
38. Stoney, 2005
39. Trotter, 1991
40. Welsh et al., 1992
41. Wilkinson, et al, 2001
42. Yang et al. 2013
43. Zhao et al. 2013

Chapter 7

Data Modeling and Visualization in GIS and Remote Sensing

CONTENTS

The chapter basically focuses on the process of data modeling and visualization that takes place in GIS and remote sensing. At the beginning of the chapter, it provides a detailed introduction about the data models used in GIS in order to obtain the desired output. Next, the chapter puts light on the integration of gradient modeling and remote sensing. In GIS, there is statistical data visualization that takes place in order to visualize and analyze the data. One topic named 'immersive 3D visualization of remote sensing data' is the topic that provides insights about the methods and processes or even software that are widely used in order to produce three-dimensional visualization.

7.1. INTRODUCTION

In a number of different contexts, the term 'modeling' can be used considering the world of GIS. Therefore, it would be wise if the start begins with an effort in order to clarify the meaning of the term. There can be particularly two meanings. The first and the foremost say, 'a data model can be defined as a set of expectations related to the data. This means that it is a template into which data required for a specific application can be filled in.'

Let's take an example. A very simple example of a date model is a table. It is important to know the way in which the tables are used most of the time in GIS, the rows of table in correspondence to a group or class of features in real – world which include countries, trees or lakes. Also, it can said to be the columns in correspondence to the various characteristics of the features that can be termed as the attributes.

The second definition says, 'a model, that does not have the data qualification, is considered as a representation of one or more than one processes which are believed to occur in the real world or can say, having the belief of the way in which the world works. A model can be considered as a computer program that takes a digital representation of one or more than one aspect of the real world and further transforms the in order to create a new representation.

If the input and output both correspond to the same point in time, then the models are referred to as static. If in case, the output represents a later point in time in comparison to the input then the models are referred to as dynamic. In all these models, the common point is the operation of GIS in a number of different stages, whether they are being used for creating complex indicators from input or for representing the time steps in the operation of a dynamic process.

Most of the time, the static models take the form of indicators further combining different inputs for creating a useful output. An example can be of the USLE that is the Universal Soil Loss Equation. This combines the layers of mapped information related to soil quality, slope, agricultural practices and some other properties in order to estimate the amount of soil that will be lost to erosion from a unit area in a unit time (Wischmeier and Smith, 1978).

GIS tends to store the data from which the user can draw the desired view in order to suit a particular purpose. This data is referred to as geographic data. Basically, there are two types of data in GIS such as spatial data and non-spatial data.

The non-spatial data include the information which is related to the features such as the name of roads, forests, schools, related to the population or census data for a particularly given region, etc. The non-spatial or attribute data is the data that qualifies the spatial data. It helps in describing some of the aspects of the spatial data which is not specified by its geometry only.

The integration of the above two types of data is done by using GIS. It allows users to derive new data for the purpose of planning. In such a way, the spatial models are important in which the information is represented affects the type of analysis which can be performed and also, the type of graphic display that can be performed and the type of analysis which can further be obtained.

While considering the GIS, it can be said that there is a major difference between vector GIS and raster GIS. These are the two approaches for spatial data processing. These approaches are to be found most of the time in the same GIS package. These approaches tend to reflect on two different methods of spatial modeling. The former focuses on the discrete objects which are to be described and the latter is concerned primarily with the recording of that which is to be found at the pre-determined set of locations which can be grid cells or points.

The remotely sensed images are analyzed for environmental as well as agricultural monitoring. This along with the map update represents the main topic of remote sensing. Along with the conventional visual or infrared camera, the SAR that is Synthetic Aperture Radar sensor becomes more and more popular and the reason behind this is that it works at night and penetrates the clouds.

Now, in today's scenario, the evaluation of digital images of the same observation is done in parallel while displaying them on different monitors along with a map. Because of the different image scales and orientations, it becomes difficult for the human interpreter to explain the image content or to even recognize the corresponding structures. The combination of the multi – sensor data is envisaged on a single screen in order to improve the evaluation facilities.

The explanation of the unknown structures and the detection of temporal changes are eased by the simultaneous visualization. The reason behind this is that the comparison of image data is done directly with other images and the map. It is important to have three-dimensional real- time visualization in order to give a spatial impression of the observed terrain.

This further allows the human interpreter to explore as well as evaluate the multi – sensor image data in a virtual landscape. Vector data of a GIS is made accessible in order to ease the image interpretation.

In the past time, there have been different 3D visualization systems that have been developed for the geographical data. These systems tend to differ in several aspects such as the used datasets, their application, visualization techniques, etc. For example, the system ViRXIS serves as a base for a VR based Tourist Information System or a VR based Geographic Information System (that is VirGIS) respectively.

It uses a DEM that is Digital Elevation Model and the raster data such as satellite imagery or scanned maps. The elevation model is represented in different Levels of Details (that is LOD) in order to guarantee a high visualization performance.

There was a system that was developed by Nuesch (in the year 1995). This system was specifically for the real- time landscape visualization based on Landsat TM images and a DEM. In addition to this, there are vector data such as roads and buildings which are inserted into the landscape model by further modifying the TIN that is Triangulated Irregular Network further representing the surface.

Besides the LOD management, the use of geo-typical texture increases the display performance. This geo-typical texture is taken from a database rather than the photo texture from the image data. McKeown (in the year 1994) described a similar approach. Along with the satellite imagery, they tend to use aerial images. The given vector data set contain information related to the buildings and roads. This data set is extended by automatically extracted features, the integration of which has been done in the TIN.

The visualization can be defined as the mental as well as the physical process of representing a set or sets of data graphically and using the very broadband channel of human visual information flow. It also uses the human pattern recognition abilities and the human access to a structure that is rich and cognitive in order to impart hypotheses, theories and explanations which pertain to the causal processes which are inherent in the data.

7.2. DATA MODELS FOR GIS

It is important to determine the way to best represent the geographic space in order to visualize the natural phenomena. Data models are said to be a set of rule or constructs which are used for describing and representing the aspects of the real world in a computer. There are two primary data models that are available so as to complete this task. They are named as raster data models and vector data models.

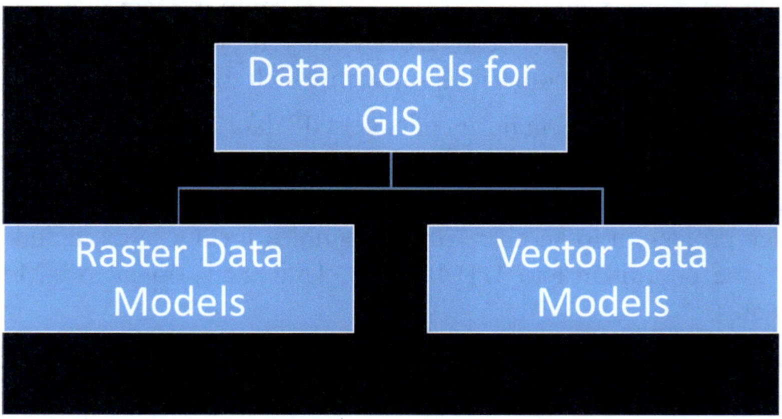

Figure 7.1. Types of data models for GIS.

7.2.1. Raster Data Models

The use of raster data model can be found widely in applications ranging far beyond geographic information systems. It is much likely that everyone is already very familiar with this data model if anyone has any such experience with digital photographs. The ubiquitous BMP, JPEG and TIFF file formats (among others) are based on the raster data model.

Let's take an example of viewing a digital image. If the image is deeply zoomed into, then it can be seen that it contains an array of tiny square pixels or picture elements. Each of these pixels is uniquely colored pixels. When these are viewed as a whole, then it combines in order to form a coherent image.

Moreover, it can be said that LCD that is Liquid Crystal Display computer monitors are based on the raster technology, because of the fact that they contain a set number of rows and columns of pixels. It can further be noted that the base of this technology predates computers as well as digital cameras by nearly a decade or more.

There was a neo-impressionist artist named Georges Seurat. In the period of 1880s, he developed a painting technique which was referred to as pointillism. This relies similarly on the amassing of small and monochromatic dots of ink which in combination form a larger image.

The raster data model contains rows and columns which are of equal -sized pixels. These pixels are interconnected to form a planar surface. It can be said that these pixels are used as building blocks in order to create points, areas, lines, networks and surfaces. Despite the fact that the pixels can be triangles, hexagons or can be even octagons, the square pixels tend to represent the simplest geometric form with which to work.

Accordingly, the vast majority of raster GIS data that is available are built on the square pixel. Typically, these squares are reformed into rectangles having different dimensions if in the case; the data model is transformed from one projection to another. Some examples for the same include from State Plane coordinates to UTM that is Universal Transverse Mercator coordinates.

The raster data model is referred to as a grid-based system. This is because of the fact that the raster data model relies on a uniform series of square pixels. Typically, each grid locale is assigned with a single data value. In the raster, each cell carries a single value and this tends to represent the characteristic of the spatial phenomenon at a location that can be denoted by its row and column.

For that cell value, the data type can either be an integer or be floating-point. In an alternative context, the raster graphic can reference a database management system. The system in which open-ended attribute tables can be used in order to associate multiple data values to each pixel.

The advancements or modifications have been made in computer technology and such advancements have made this second methodology increasingly feasible due to the fact that the large datasets are no longer constrained by the issues related to the computer storage as they were previously.

Within a given pixel, the raster model will average all values in order to yield a single value. Hence, it can be said that the more area is covered under per pixel, the less accurate the related data values will be. The area that is covered by each pixel determines the spatial resolution of the raster model (from which it is derived). Particularly, the resolution can be determined by measuring one side of the square pixel.

A raster model having pixels equal to 10 m by 10 m or 100 sq meters in the real world, it would be said to have a spatial resolution of 10 m. A raster model having pixels 1 km by 1 km or 1 sq. km in the real world then it would be said to have a spatial resolution of 1 km.

It is important to take care while determining the resolution of a raster because of the fact that using an overlay coarse pixel resolution will cause a loss of information. On the other hand, if overly fine pixel resolution is used then it will result in considerable increases in file size and computer processing requirements during the display or analysis.

An effective pixel resolution will consider both the map scale as well as the minimum mapping unit of the other GIS data. Considering the case of raster graphics having a coarse spatial resolution, it is not necessary that the data values related to the specific locations are explicit in the raster model.

Let's take an example. If the location of the telephone poles were to be mapped on a coarse raster graphic, then it would be clear that the whole cell would not be filled by the pole. Instead of this, the pole would be assumed to be located somewhere within that cell; typically the position can be at the center.

There should be a number of different properties that should be there in imagery while employing the raster data model. One of them is that each pixel is required to hold at least one value, even though that data value is zero. Moreover, for a given pixel, if there is no data present, then it is important to assign a data value placeholder to this grid cell.

Most of the time, an arbitrary and readily identifiable value will be assigned (for example, -9999) for the pixels for which there is no data value. Another property can be that a cell can hold any alphanumeric index which represents an attribute. Considering the case of quantitative datasets, the attribute assignation is fairly straightforward.

An example can be taken as if a raster image denotes elevation, then for each pixel, the data values would be some indication of elevation in feet or meters (usually). Considering the case of qualitative datasets, the data values are indices. Some pre-determined translational rule is referred to by these data values.

Considering the case of a land – use or land – cover raster graphic, there can be an application of the following rule:

- Grassland
- Agricultural
- Distributed and so on.

The third property of the raster model data can be that the points and lines move towards the center of the cell. Similar to one's expectation, if a 1 km resolution raster image contains a river or stream, and then it would not be clear to determine the location of the actual waterway within the pixel of the river. Thus, there is a general assumption that is taken into consideration, that is, all zero – dimensional (point) and one – dimensional (line) features will be located towards the center of the cell.

As a corollary, for any line feature, the minimum width should necessarily be one cell without considering the actual width of the feature. If it is not, then the representation of the feature will not be done in the image and will thus be assumed to be absent.

There are a number of different methods that exist in order to encode the raster data from scratch. Among these, there are three models which are as follows:

- Run-length raster encoding

This method named 'run-length raster encoding' encodes cell values in runs of similarly valued pixels. This can further lead to a highly compressed image file. This run-length encoding method is said to be useful in the situations where there are similar values in the large groups of neighboring pixels (for example, discrete datasets which include land use or land cover or habitat suitability) and is not much or less useful where there is variety in the neighboring pixel values (for example, continuous datasets which include elevation or sea - surface temperatures).

- Cell-by-cell raster encoding

This method named 'cell-by-cell raster encoding' is a minimally intensive method. This method helps in encoding a raster by further creating records for each cell value by row and column. This method could be considered as a large spreadsheet in which each cell of the spreadsheet tends to represent a pixel in the raster image. Also, this method is referred to as "*exhaustive enumeration*."

- Quad-tree raster encoding

This method named 'quad-tree raster encoding' divides a raster into a hierarchy of quadrants. These quadrants are subdivided on the basis of similarly valued pixels. When a quadrant is made entirely from cells of the same value, then the division of the raster stops. A quadrant which cannot be sub-divided is referred to as a "*leaf node*."

7.2.2. Vector Data Models

The Vector data model is said to be in contrast to the raster data model. In this model, space is not quantized into discrete grid cells as there is in the raster model.

Points and their associated X, Y coordinate pairs are used in the vector data models in order to represent the vertices of spatial features as if they were being drawn on a map by using the hands (Aronoff, 1989).

The storage of the data attributes of these features is done in a separate DBMS that is a database management system. There is a connection between the spatial information and the attribute information for these models and these connections are formed via a simple identification number that is provided to each feature in a map.

There are three fundamental vector types that exist in geographic information systems (GISs). These types include points, lines, and polygons. Points are said to be the zero - dimensional objects which contain only a single coordinate pair.

Typically, the points are used for the purpose of modeling the singular, discrete features which include buildings, power poles, wells, sample locations, and many more.

There is the property of the location that is found in points. This is the only property that is found in points. The other types of point features are the node and the vertex. Particularly, a point is said to be a stand-alone feature, whereas a node is said to be a topological junction that is representative of a common X, Y coordinate pair between the intersecting lines or polygons.

The vertices are defined as each bend that are there along a line or polygon feature which is not the intersection of lines or polygons.

Points can be spatially connected in order to form more complex features. The lines are defined as the one-dimensional features which contain multiple as well as explicitly connected points.

Lines are used for the purpose of representing the linear features which include roads, faults, streams, boundaries, and much more. There is a property of length which is found in lines. The lines which directly connect two nodes are referred to as chains, segments, edges, or arcs.

Polygons are defined as the two-dimensional features which are created by a number of lines that tend to loop back for creating a "closed" feature. Considering the case of polygons, the first coordinate pair (point) on the first line segment is the similar one as the last coordinate pair on the last line segment.

Polygons are used for the purpose of representing the features which include city boundaries, lakes, geologic formations, vegetation communities, soil associations, and much more. Polygons contain the properties of area and perimeter. Polygons are also referred to as areas.

7.3. SATELLITE IMAGERY AND AERIAL PHOTOGRAPHY

There is a wide variety of satellite imagery and aerial photography available to be used in geographic information systems (GIS). Within the GIS, these products are used differently, although being graphically raster. The information provided by the Satellite imagery and aerial photography is based on some context for GIS.

They are often being used for conducting heads-up digitizing wherein the features from the image are converted into vector datasets.

7.3.1. Satellite Imagery

Around the globe, the use of remotely sensed satellites is being continually sent into space by various private companies and public agencies. It has made remotely sensed satellite imaging very common. There are several uses of satellites such as communication, weather, civilian Earth observation, military and many more.

Figure 7.2. Neizvestnyye Islands - Sentinel-2 satellite imagery.

Source: Image by Wikimedia Commons

As of now, there are more than 4000 satellites out of which 3000 are from Russia and the United States of America that have been sent to space. The satellites have to maintain their inclinations, synchrony, orbital centers, different attitudes and this allows them to capture the image of the surface of the earth with various features and processes.

That can be an active satellite or a passive satellite. The use of remote sensors in satellites detects the responses that are reflected from the object. These are radiated from the energy sources that are generated artificially. These are the active satellites.

For example, light waves emitted from laser sensors, radio waves from radars and sound waves emitted from Sonar sensors. In all the cases a common thing is the sensor emits the signal and then calculates the time in which the returned signal bounce back from some remote feature.

The distance from the sensor to the feature can be calculated from the delay in the time when the original admission took place to the return, knowing the speed of the emitted signal.

On the other hand, passive satellites make the use of a sensor that detects the reflection or emission of electromagnetic radiation from natural sources. The natural source is the energy source from the Sun. However, other sources can also be images such as geothermal activity and magnetism.

The examples of a passive satellite are: capturing an image with a flash -enabled camera would be an example of active remote sensing while capturing the image without using the flash would be an example of passive remote sensing (i.e., relying on ambient light to illuminate the scene) would be passive remote sensing.

Resolution is the main factor that determines the quality and quantity of the satellite imagery. Any remote sensor can have four types of resolution (Campbell 2002). The spatial resolution of a satellite image directly represents the ground coverage for each of the pixels shown in the image. This was also previously discussed in the raster, data model section. A 10 m resolution image produced by the satellite imagery gives the ground coverage for each of the pixels and that is 10 m by 10 m or 100 square meters on the ground.

The Instantaneous Field Of View (IFOV) of the sensor determines the spatial resolution. The instantaneous field of view is basically the area of ground through which the sensor receives the electromagnetic radiation signal. It is determined by the height and angle of the imaging platform.

The ability of the sensor to resolve wavelength intervals which are also called bands, within the electromagnetic spectrum is denoted by Spectral resolution. The interval size of the wavelengths and the number of intervals being scanned is the factor which is determined by the spectral resolution.

Within the spectrum, there are two types of sensors, multispectral and hyperspectral sensors that can resolve a multitude of wavelengths intervals. There are some examples of the satellite images, IKONOS satellite resolves images for bands at the blue (445–516 nm), red (632–98 nm), green (506–95 nm), and near-infrared (757–853 nm) wavelength intervals on its 4-meter multispectral sensor.

The temporal resolution is the amount of time between each image collection period. The repeat cycle of the satellite's orbit determines the time between each image collection. The temporal resolution can be a true-nadir or off-nadir. The areas that are located directly beneath the sensors are the areas in true-nadir. On the other hand, the areas with the images taken oblique are the areas considered in off-nadir.

The temporal resolution of the IKONOS satellite is considered to be 3 to 5 days in the case of off-nadir imaging. While, it is 144 days which require for true-nadir imaging. The radiometric resolution is the fourth and final type of resolution that refers to the sensitivity of the sensors that vary in brightness. This type of resolution specifically denotes the number of grayscale levels which are imaged by the sensor.

If it is seen typically, the radiometric values available for an 8-bit sensor yield the values that range from 0–255 as 256 unique values or as 2^8 values. Similarly, for 11-bit, it is 0–2,047, 12-bit, 0–4,095, and 16-bit, 0–63,535. One of the examples is Landsat-7 that maintains an 8-bit resolution for its bands. Therefore, it can record values for each pixel that range from 0 to 255.

There is a trade -off state between the different types of resolution. This is due to the technical constraints that the satellite remote sensing systems have. If there is an improvement in one type of resolution, that necessarily needs a change (reduction) in the other types of resolution. For example, if there is an increase in the spatial resolution, it necessitates the decrease in spectral resolution, and vice versa is also true.

In the case of geostationary satellites (that circle the earth once each day and is proximal to the equator), there is a high temporal resolution and low spatial resolution. On the other hand, in the case of the sun-synchronous satellites (which synchronize a near-polar orbit of the sensor with the help of sun's illumination), it yields a low temporal resolution and provides a high spatial resolution.

However, there have been numerous technological advances that have taken place. This has improved the various resolutions of an image. However, care must be given to make sure that the imagery is adequate that represents a geospatial feature that is most important to study.

7.3.2. Aerial Photography

There is a vast source of information in the aerial photography for use in the GIS. It is similar to the satellite imagery in its vastness. The aerial photographs use the hardware platforms such as helicopter, balloons, rockets, airplanes and many more. The images taken from aerial photography are from the visible spectrum and the sensors measure the bands within the non-visible spectrum (e.g., ultraviolet, infrared, near-infrared). However, it can also be fixed to the aerial sources.

Just like satellite imagery, aerial photography can also be active or passive. The images can be taken from the vertical or oblique angles. The aerial photographs are to be taken carefully as the images captured from the sensors are similar to the images taken from cameras by using the lenses.

The image becomes more pronounced as a person moves away from the center of the image due to the lenses. The lenses add curvature to the image and hence the image becomes more noticeable. Another source of error in aerial photography may be relief displacement. This error is due to the three-dimensional aspect of the image features.

Figure 7.3. Aerial Photography of Colosseum – Rome, Italy.

Source: Image by Flickr

The vertical objects in this photography re seen leaning away from the center point of an aerial photograph. This type of error can be understood by imagining that a smokestack would look like a doughnut if the camera is just above the feature. In another case, when the camera is slightly at the side of the smokestack, it would give another side of the image feature.

This error is mostly noticed in the case of the images taken of the trees, multistory buildings. Therefore, it is not suitable to capture images of the taller features. Orthophotos are those images that are corrected geometrically to make it curvature free and remove the terrain-induced errors from the image. The orthophotos are captured vertically.

The most common product of the orthophoto is the Digital Ortho Quarter Quadrangle (DOQQ). The digital ortho quarter quadrangle was first made available through the US Geological Survey (USGS). USGS started producing these types of images from the library of National Aerial Photography Program photos on the scale of 1:40,000

The images can be made available in the radiometric resolution of 1-meter spatial resolution and 8-bit. It can be either in grayscale or color.

The name and the features of this image type are similar. The images from the USGS cover a quarter of a USGS 7.5 minutes quadrangle and this is equal to the 25 square mile area approximately. The photos also include the additional edge of around 50 to 300-meter which makes the users to mosaic many DOQQs into a single and continuous image. These DOQQs images are ideal to be used in the Geographical information system (GIS) for the purpose of background display information, heads-up digitizing and data editing.

7.4. INTEGRATION OF GRADIENT MODELLING AND REMOTE SENSING

In order to develop better ecological maps, there are some recent mapping and classification of image efforts that illustrate the power of formally melding environmental information with that of satellite imagery. In order to construct an effective tool for exploring oyster population dynamics, Michener and others (1992) combined field data, GIS, and spatial statistics.

The link of the environmental gradients which is derived from the climate models and digital maps with the regional plot data is demonstrated by Ohmann (1996). This was done in order to derive information that is relevant to forest planning and policy. For characterizing the woody species composition in Oregon, both Ohmann and Spies (1998), made use of the same methods for identifying the regional gradients from extensive field data.

At a regional scale, Ohmann and Spies developed a conceptual model of species- environment relations. This model can help in accurately map the forest species when used with remote sensing. In the North Cascades Mountains in Washington, U.S.A, the gradient analysis was linked with the spectral data and the prediction was made with regards to forest species distributions.

This study is unique in its own way and it makes it different from many other studies explained above. Firstly, there is an integration of the remote sensing with the comprehensive gradient-based field inventory sampling system. The mapping process becomes easier with ecosystem simulation.

Secondly, the studies that were presented earlier support in the creation of only one product i.e., a vegetation map. On the other hand, this study is used to map a wide variety of ecological attributes.

Third, the maps were created from LEIS and are made up of the probability surfaces. This refers to the maps that represent the probability of possessing a specific characteristic in any area of the map.

In contrast to the other mapping efforts being done, where the final data output was presented as nominal or ordinal categories. This enables the end-users of the output maps for determining their own strategies from LEIS and for classification of the characteristic of ecology. Therefore, for a wide variety of data layers, it helps in maximizing the utility of the final spatial data.

The fourth and the different aspect of this study is that there are many environmental gradients for the prediction of ecosystem characteristics. Therefore, it is different from other studies. Among this, there are many eco-physiological direct or functional gradients.

This study contrasts with that of Kessell (1979), where there was the usage of only seven indirect gradients for predicting vegetation and fuels in Glacier National Park.

Fifth, the gradient modeling approach has integrated the primary eco-physiological gradients which dictate ecosystem properties such as the productivity, leaf area index, and evapotranspiration. This has formally integrated with that of the modeling approach.

Sixth, with the conventional remote sensing, it has allowed the control of the ecosystems with a more fundamental spatial characterization of the gradients. This study has evolved a new remote sensing technology and the simulation of ecosystem software with field sampling. Last but not the least, this study has implemented the LEIS on any land area with the help of methods and protocols.

7.5. STATISTICAL DATA VISUALIZATION IN GEOGRAPHICAL INFORMATION SYSTEMS

Cartograms are considered as a unique representation for the geographical space. Cartogram is a type of graphic tools that helps in describing the attributes of geographic objects. Mapping requirements basically includes the preservation of orientation contiguity, shape, and data that are having suitable variation.

It totally depends on the map reader, how well they recognize the shapes of the internal enumeration unit for successful communication. It also depends on the accuracy of estimating these areas and effective fable design.

As cartogram is seen not to depict the geographic space, but rather it is observed to have changed in the size of objects which depends on a certain attribute, therefore cartogram is not a true map. Cartogram is said to vary on their degree in which there is basically change in the geographic space. Therefore, some of them appear very similar to a map, but some do not look like a map.

Basically, there are three types of cartograms, each of these cartograms are consider having a different way of showing attributes of the geographic objects. These objects are namely: non-contiguous, contiguous, and Dorling cartograms.

A non-contiguous cartogram is the easiest and the simplest type of cartogram. In a non-contiguous cartogram, it is considered that the geographic objects do not need to maintain the connectivity with their adjacent objects. This type of connectivity is called a topology. In order to release the objects from the vicinity, this often makes the objects grow or even get smaller in size, and still they are able to maintain their shapes.

In a contiguous cartogram, it is believed that it maintains its true topology. In this, the object even remains connected with each other, but it is also seen that this even causes great distortion in the shape. In this case, it is important that the cartographer makes the object with the appropriate size in order to represent the attribute value. But with that it also important to try to maintain the shape of the objects as much as possible, because with this the cartogram can be easily interpreted.

A Dorling cartogram is believed that it neither maintain their shape topology not they maintain object centroids. It has been proven to be a very effective cartogram method. Dorling cartogram was named after its inventor, Danny Dorling of the University of Leeds.

Therefore, in order to create a Dorling cartogram, they did not expand or distort the object, instead of that, the cartographer replaces the objects with the uniform shape which is usually a circle of the appropriate size.

7.6. VISUALIZATION OF SPATIAL DATA TO SUPPORT FIELD DATA ACQUISITION

One of the main reasons behind the quick adoption of visual GIS technology is that the visualization comes as naturally to the users (Buttenfield, 1996). The procedure of digital outdoor data captures often requires the application of visualization, because it is used as a means to recognize and interpret the visual pattern.

Such kind of pattern occurs on the maps and other images are present on the pen computer screen. This kind of pattern can be even compared straightway with the outdoor situation. Using this sense, the visualization often seems to serve the purpose of clarifying and even enhancing the understanding about the conditions of the real world (Beard et al., 1994).

It is being noticed that the user basically "sees" the object of interest which in both the forms, either naturally or symbolically present on the digital graphic display. This helps by contributing to an easier process that is used for matching the symbol dimensions which are used in maps, directly with the phenomenon dimensions which are basically occurring in nature. These are considered as a basic requirement of a cartographic language (MacEachren, 1994).

There are few examples that show the usefulness of such tools which exist in various forms:

- Modification and improvement of various kinds of ecological maps. For example, geological maps, vegetation cover maps and soil maps.
- Updating land use maps.
- Improvement and evaluation of thematic maps and other graphics.
- Comparison of historic photographs and maps with the current field situation.

- Updating topographic maps.

Mobile GIS which is a field- based, is also being used for environmental characterization, decision support and modeling. There are several advantages that lie in the interactive development of sampling strategies that are through visualization, presence of on the spot environmental modeling and feedback that is received through the field- based verification (Carver et al., 1996).

Navigation and orientation that are present in the field are considered as an additional issue that must be supported by mobile GIS. Maps are designed in such ways that they help to communicate with a special message to a user (Wood, 1994).

This simply means that the use of topographic and thematic maps in the field can help in contributing towards a more comprehensive understanding of the outdoor situation.

In order to support such goals, GIS those are field- based is believed to provide topographic, cartographic and geo-referenced information. It is considered that adequate support that is given to the user is basically to provide the raster maps. These are the maps on which the user can be able to define the new objects, select those objects, and later on add the attributive information to them.

Specifically, for the users, the digital raster maps with the GPS links help them and support them to orient and navigate. The raster maps are believed to get overlaid with the various vector data. For example, social topographical information's like streams, parcel boundaries and roads.

These functionalities often guarantee adequate support for capturing the data. Given below are a few examples of projects relayed to same:

- Collection of parcel-related information in agriculture, this project is based on the polygons for which information related to land use, size of land, and many others are captured.
- Forest inventories, in this project polygon, are being defined and there is a collection of comprehensive data of each tree that is present in that area are collected.
- River habitat surveying, in this survey the river sections are being defined and the details are often provided by using the habitat data.

Such visual support is considered to be relatively simple which is used to realize within a less or more closed system. Although, in many cases, the field workers who want to use the data that is available on various sources.

For example, they often need data that is available from the stationary GIS in an office, it can be also scanned from raster maps, vector geometries that are gathered from the official topographic services and other information.

The open flow which is of different raster and vector formats that are between the GIS is considered as an issue that is currently under the intensive discussion and development that takes place within the framework of the OpenGIS initiative (McKee and Kuhn, 1997; OpenGIS, 1998a).

Field- based GIS is needed to support, which is therefore, is the simple Features Specification of the OpenGIS Consortium (OpenGIS, 1998b). The specification of interoperable tools for the filed based GIS is becoming an important branch of research and development.

Interoperability is considered to play a significant role specifically in the provision of visual add-on information. For ecological purposes, for example, assessments and classification must be carried out.

Field workers are especially those persons who are not having the required experience in ecological mapping. They even do not have sophisticated knowledge that is even required to perform adequately such a complex task.

Whereas, it is believed that the diagram, images and photographs that are in 3D visualization can help in representing the support in each of these situations. These approaches help in bridging the gap to the current initiatives, looking at the multimedia technology potential for the display and visualization of spatial data (Masser and SalgeÂ, 1997).

The visual capabilities that are being mentioned are further combined with the knowledge- based function which basically aims at achieving the high quality of data (Pundt and Kuhn, 1998), and with that make the field -based GIS indispensable and comprehensive components of environment information system.

7.7. IMMERSIVE 3D VISUALIZATION OF REMOTE SENSING DATA

It is often believed that a 3D site model generally includes a description of the terrain, about the streets, about vegetation and building of a particular concerned area which needs to be built-up. Even though, for several applications, there is a requirement of additional information which mainly includes man-made objects and buildings which are considered as an important factor 3D site model.

For instance, for the visual reality application, there is a realistic representation that can be acquired only when there is a texture of the ground, facades, the roof is available. There are also some other significant details like trees, fences, footpaths, etc.

The basic purpose which was given behind this work is the technology of the 3D site model which is being generated in the PC-level computers. In this technology, the digital terrain models are present on large territories that are created by using the space images, man- made objects and terrain models which are created by using a large scale of aerial images.

7.7.1. Image Representation

The data that is acquired from the various sources such as satellites, aerial -based images are considered to be of two dimensional in nature. The two-dimensional images are considered as a raster image where it consists of pixels. Here the pixel is considered as the smallest element. The images are formed by using these pixels. Therefore, the data that is acquired from the various aerial images, will have the two- dimensional representation.

7.7.2. Two Dimensional Picture of an Image

Figure 7.4. Two-dimensional image of space.

Source: Image by Solar system exploration research virtual institute-NASA

The image which is shown in the above image is in two -dimensional pictures. The image has a Gray value that is shown in the above image. The images will be having some different colors. The various color of the image is considered to be RBG, where R is for red, G is for Green and B is for blue indexed color mode.

The two -dimensional image can be of 8 bit, 16 bit, or 24 bit. Where 8 -bit image occupies 1 byte, 16 bits occupies 2 bytes and 24 bits image occupies 3 bytes.

7.7.3. Three Dimensional Picture of an Image

For the purpose of representing the Three Dimensional pictures of the image, it is important that one needs the two- dimensional picture of the height and image information that is basically constructed with the elevation of the image.

Therefore, the elevation of an image is being constructed with the models of the Digital Elevation Models and Digital Terrain Models. It is being observed that in the DEM, the elevation is being computed without the surface information. But it is being taken into account in the DTM surface models.

7.7.4. Remotely Sensed Satellite Imagery

Remote sensing is said to be the acquisition of information related to an object or phenomenon without making physical contact with the object. Considering the contemporary parlance, the remote sensing refers to the use of aerial sensor technologies in order to detect and classify the objects present on Earth (both on the surface and in the atmosphere as well as oceans) through the transmitted signals such as electromagnetic radiation which are emitted from aircraft or satellites.

Figure 7.5. Remote sensing imagery of a changing Arctic.

Source: Image by U.S. National Park Service

There are a number of purposes which are served by these satellites in the fields of meteorology, geology, agriculture, forestry, biodiversity conservation, landscape, education, regional planning, intelligence and warfare. There are two key categories of remote sensing which exist. They are named as passive remote sensing and active remote sensing.

Passive Sensors have the ability to detect natural radiation. These natural radiations emit or reflect by the object or the surrounding area which is being observed.

Reflected Sunlight is said to be the most common source of radiation that is measured by passive sensors. There are some examples of passive remote sensors that include film photography, infrared charge – coupled devices and the radiometers.

In contrast to this, the active remote sensing emits energy that scans objects and area. After this, the radiation reflected or backscattered from the target is detected and measured by a sensor.

One example of the active remote sensing is RADAR. In this, the time spent between the emission and return is measured. This further establishes the location, speed, height and direction of an object.

The human has gained ability due to the remote sensing technology to acquire data on and from hazardous as well as inaccessible areas.

The applications of remote sensing are being used in order to monitor the deforestation in areas including the Amazon basin or the glacial features in Arctic and Antarctic regions and furthermore, collecting the information by means of waves in coastal as well as ocean depths.

During the period of the cold war, the Military used the data collected related to dangerous border areas. Also, the remote sensing replaces costly as well as slow data collection on the ground; this further ensures that there are no disturbances in the areas or objects during the process. The quality of remote sensing data further relies on its spatial, radiometric, spectral and temporal resolutions.

The size of a pixel that is recorded in a raster image is referred to as Spatial Resolution. The pixels match up to the square area. This tends to range in side length from 1 to 1,000 meters (that is equal to 3.3 to 3,300 ft).

The wavelength width of the different frequency bands that are being recorded is referred to as Spectral Resolution. In general terms, this is related to the number of frequency bands which is recorded by the platform.

The number of different intensities of radiation in which the sensor is able to differentiate between is referred to as Radiometric Resolution. Usually, it ranges between 8 to 14 bits, which tends to correspond to 256 levels of the grayscale and is up to 16,384 intensities or "shades" of color, in each of the band. Also, it depends on the instrument's noise.

7.7.5. Open Scene Graph

For the purpose of the development of high- performance graphics applications which is via games, flight simulators, scientific visualization and virtual reality. Accordingly, the open -source library is the open- source library and cross platform graphics toolkit.

It is basically based on the concept of a Scene Graph that is considered that it provides an object which is an oriented framework on the top of the OpenGL. This basically gives the developer a scope that focuses on many additional utilities for the fast development of the graphic application as well as sets the developer free from the implementing and optimizing the low level of graphics calls.

7.8. CONCLUSION

In its true sense of the word, the GIS that is Geographical Information System has never been used in order to produce the maps. The true maps are being produced tend to use the output obtained by GIS which is presently being produced with the auxiliary software such as desktop which publishes packages or computer cartography software. Therefore, the GIS is developing competent cartographic display packages.

Technically, and not technological, the research considers the animation, 3-D images and space-time representations while considering the visualization tools for inquiry. The conclusion that can be drawn from the above chapter is that there has a lot of development that has took place in GIS and remote sensing and so, there has been advancement in the process of data modeling and visualization in both the processes.

REFERENCES

1. GOODCHILD, M., n.d. *GIS And Modeling Overview*. [ebook] p.1. Available at: <http://www.geog.ucsb.edu/~good/papers/414.pdf> [Accessed 24 April 2020].
2. Growe, S., Schulze, P. and Tonjes, R., n.d. *3D Visualization And Evaluation Of Remote Sensing Data*. [online] Citeseerx.ist.psu.edu. Available at: <http://citeseerx.ist.psu.edu/viewdoc/download?doi=10.1.1.45.1644&rep=rep1&type=pdf> [Accessed 24 April 2020].
3. Keane, R., Rollins, M., McNicoll, C. and Parsons, R., 2002. *Integrating Ecosystem Sampling, Gradient Modeling, Remote Sensing, And Ecosystem Simulation To Create Spatially Explicit Landscape Inventories*. [ebook] p.2. Available at: <https://www.fs.fed.us/rm/pubs/rmrs_gtr092.pdf> [Accessed 24 April 2020].
4. Pundt, H. and BrinkkoÈtter-Runde, K., 1998. *Visualization Of Spatial Data For Field Based GIS*. [ebook] p.2. Available at: <http://web.mit.edu/envit/www/pdf/field-basedGIS.pdf> [Accessed 24 April 2020].
5. Rautji, S., Gaur, D. and Khare, K., 2013. *IMMERSIVE 3D VISUALIZATION OF REMOTE SENSING DATA*. [ebook] p.2. Available at: <https://www.researchgate.net/publication/274173757_Immersive_3D_Visualization_of_Remote_Sensing_Data/link/56880d1608ae19758398e86f/download> [Accessed 24 April 2020].
6. REVEIU, A. and DÂRDALĂ, M., 2011. *Techniques For Statistical Data Visualization In GIS*. [ebook] p.1. Available at: <http://revistaie.ase.ro/content/59/06%20-%20Reveiu,%20Dardala.pdf> [Accessed 24 April 2020].
7. Saylordotorg.github.io. n.d. *Data Models For GIS*. [online] Available at: <https://saylordotorg.github.io/text_essentials-of-geographic-information-systems/s08-data-models-for-gis.html> [Accessed 24 April 2020].
8. vanDerMerwe, F., 2003. *GIS, MAPS AND VISUALISATION*. [ebook] p.2. Available at: <https://www.researchgate.net/publication/228990989_GIS_Maps_and_Visualisation/link/5513df220cf23203199cc734/download> [Accessed 24 April 2020].
9. Wischmeier and Smith, 1978
10. Nuesch, 1995

11. McKeown, 1994
12. Aronoff, 1989
13. Campbell 2002
14. Michener and others, 1992
15. Ohmann and Spies, 1998
16. Kessell, 1979
17. Buttenfield, 1996
18. MacEachren, 1994
19. Beard et al., 1994
20. Carver et al., 1996
21. Wood, 1994
22. McKee and Kuhn, 1997
23. OpenGIS, 1998
24. Masser and SalgeÂ, 1997
25. Pundt and Kuhn, 1998

Chapter 8

Future Challenges and Perspective of GIS and Remote Sensing

CONTENTS

This chapter named 'Future Challenges and Perspective of GIS and Remote Sensing' basically provides insights about GIS and remote sensing is transforming our world. In this chapter, the readers will get to learn about the trends that are currently going on in GIS and what will the upcoming trends in the technology of GIS in the future decade. It also provides information about the future of GIS including majority of the technology that can come up in the upcoming future. The chapter also describes the future of GIST and the new technology trends going on or will come up in the future. The readers will get to learn about the future sensors which include NANO-SATS and WISP-SATS.

8.1. INTRODUCTION

There is an increase in the earth – observing satellites in space and therefore, remote sensing data are undergoing explosive growth. A large volume of remote sensing data is being produced by these satellites which eventually will give rise to complexity and diversity in characteristics of data that are high – dimensional.

Nevertheless, it is said to be extremely difficult for a number of different end -users as well as advanced user communities in order to use such traditional remote sensing data sets in an efficient manner.

Simultaneously, there will be a new challenge and that will be of data processing and distribution. It will also, become difficult enough for the user community so as to maintain the increasing volume of data and also, for building and launching the hardware into space.

There are certain crucial issues for the wide use of remote sensing data which include a large volume of data, different map projections, complex data formats, communication capability, processing time and applications of GIS.

That is why the real – time remote sensing data in a number of various formats will help the users in analyzing the future earth observing satellite systems. For better utilization of future earth – observing satellite systems, there are certain aspects that can help including the pre-processing, data format conversing, real – time data processing and open GIS applications.

There is a wide range of scope of remote sensing technology and that is from geosciences to disaster management and agriculture to urban.

In the agricultural field, the plant characteristics are studied by using crop reflectance at various stages which are included in growth. This study is then assessed in order to estimate the crop acreage and yield forecasting. Also, it studies the identification of crop as well as landholding size by using high – resolution data.

High tech machine learning tools have found their way through the conventional visual interpretation method of land use and land cover classification. The technology of remote sensing is widely used in order to plan and manage forest resources. It has been proved that the technology is useful while assessing forest biomass, forest fragmentation, deforestation, classification of forests, spatial distribution, etc.

The data obtained through Landsat, Sentinel, SPOT and IRS is used extensively in forest resource assessment and management. The data which have the very high spatial resolution is used for the purpose of classification as well as identification of built – up area in urban studies.

There are satellite sensors such as IKONOS and Quick bird which have a high spatial resolution that is less than 1 m, which are used for the purpose of cadastral mapping and urban planning.

Figure 8.1. QuickBird image from Digital Globe of Sri Lanka's Kalutara Beach before the tidal wave hit.

Source: Image by Flickr

However, the assessment of the growth of urban built-up, urban agglomeration and urban sprawl can be done by using moderate resolution data that is obtained by using Landsat, Sentinel and IRS sensors. Geologists tend to use hyperspectral images widely in order to identify the rocks and minerals. The multispectral images are also used while detecting and identifying the geological structures and rock formation by using the image interpretation keys.

Preference to the geostationary satellite data is given in order to study the extreme weather events and cyclone and the reason is that it has the ability to do constant monitoring. There are MODIS and Landsat 8 sensors which have a coastal and aerosol spectral band. These are used for aerosol optical depth. It has been proved that the remote sensing data along with geographical information system (GIS) are the most significant tools for disaster management.

Climate change analysis, weather and climate variability analysis, natural hazards mapping and vulnerability assessment are being carried out in an efficient manner by using the techniques related to remote sensing and GIS.

DEM that is the Digital elevation model is used extensively in order to analyze the slope instability and occurrence of a landslide. In addition to this, the remote sensing data are used in order to map the past and current occurrence of landslide and also, while analyzing its trend as well as patterns.

By using the remote sensing data, the earthquake mapping and flood mapping are also being done. The multispectral, hyper-spectral and DEM are used for the purpose of micro-zonation mapping of the areas which are prone to earthquake, whereas the microwave data along with DEM are used extensively for performing the assessment of flood inundation.

There is a shift that is experienced by the remote sensing technology that is from optical and multispectral to hyperspectral remote sensing. This is mainly because of the limited spectral bands. The attention of researchers has been towards the hyperspectral remote sensing or imaging spectroscopy since the last decade or so.

8.2. GIS: TRANSFORMING OUR WORLD

8.2.1. The Power of GIS

There are changes taking place in the GIS and these changes have caused changes in the way humans think and the way in which they act.

The base for this is the science of geography and also, it provides an integrative and comprehensive framework that cuts across a number of disciplines as well as organization. Also, the GIS has the power of qualitative analytics along with the geospatial visualization that is easy – to – comprehend.

GIS is improving the way of measuring and analyzing things, as well as the way in which the prediction of the future can be done, which is done by integrating geographic science into everything that is being done.

Also, it helps in providing better ways in order to plan, design and make decisions. There are thousands of important systems present around the world which have been created because of the success of the GIS community and furthermore, it is providing evidence of the huge contributions that are made by the field.

These are all the reason which have proved that the GIS is a promising technology in order to address the issues at a larger scale or a global scale or those which are present in the world. For this purpose, it is clearly required to dramatically scale up its application also and make it pervasive.

8.2.2. Scaling Up

It is important to grow dramatically the application of GIS while leveraging the current community of users. It is important to make it pervasive throughout the organizations and across society.

This scale -up is possible and in some was inevitable if the sense and experience with the pervasive technologies include the Internet and GOS is considered.

Let's take an example, the GPS has been radically simplified and this has caused a rapid transformation in the sense of place of human beings. Both the whole planet and the current place of any human can be seen along with consumer web mapping.

In a similar manner, the GIS will become pervasive further leveraging the connectivity of the web as well as cloud computing. The result thus obtained will be better decisions that reflect better understanding and eventually a more sustainable future.

8.2.3. Web GIS

In the evolution of GIS, this is said to be the next step. This step includes a new technology pattern named web GIS. Because of the release of 10.2 of

ArcGIS, it has been considered to be mapping GIS as well as complete web GIS. If tends to integrate the geospatial technology of Esri and other which means that the use of strong engineering and open standards will be done.

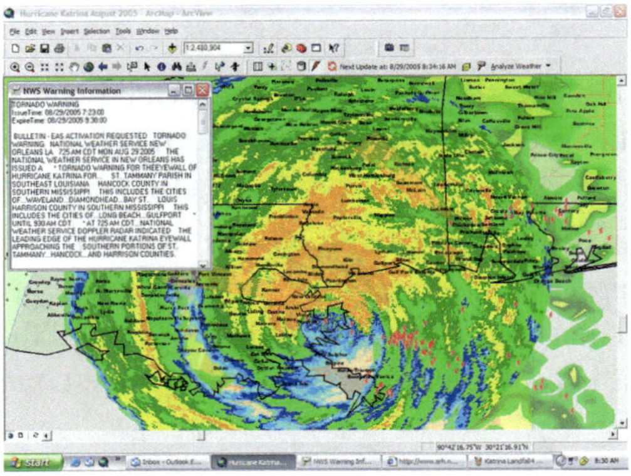

Figure 8.2. Hurricane Katrina - realtime in ArcGIS.

Source: Image by Flickr

This is the platform that takes advantage of the latest trends, which further include web and cloud technologies, faster machines, big data and pervasive devices.

Web GIS also helps in directly modeling and integration of all the geospatial data types such as from remote sensing, the sensor web, GPS, the 3D data, real – time data, crowd – sourcing data and pervasive web content.

The platform which is emerging is considered to be easy to use, accessible, more open and collaborative. Focused apps are used for making maps, performing analysis and accessing the rich as well as a living library of shared maps and geographic data.

8.2.4. Web GIS Integrates Any Data

There are a number of concepts of web GIS but one key concept if the way in which it organizes the content. Modeling of all type of information can be done by web GIS as web maps and web services.

The representation of typical geospatial data types including maps and imagery, tabular data such as spreadsheets, geo databases and enterprise databases can be done by using the web maps. Also, it tends to integrate social media as well as sensor networks along with the real – time information and the entire world of big data.

GIS is majorly all about the integration of data. In addition to this, web GIS also provides some new patterns in order to involve information sharing and collaboration, web maps and services and visually as well as analytically combining the geospatial data.

There is one more intriguing aspect of web GIS and that is it can help in breaking down the fundamental barriers which separate organization.

The ability of the web GIS environment is to fluidly integrate a number of different disciplines as well as activities and this ability gives a new framework for collaboration either the silos are professional or organizational.

There is one and the most interesting aspect of web GIS and that is that it can help in easily organizing the work. Content management capabilities are provided by web GIS for all the maps, apps and models. It also simplifies sharing these within a group or across the department as well as organizations.

8.2.5. Driving the Transformation

Web GIS can be considered as just a starting. On the other hand, it is proving it worth being an attractive platform that helps the human in scaling up and leveraging the work and knowledge of GIS professionals.

It is important for GIS professionals to understand this pattern along with its underlying technology. When these patterns are embraced then they will both leverage the existing investments and further drive the geospatial adoption across the organizations. Eventually, this pattern tends to promise to make some huge contributions to the understanding of the world.

8.3. TRENDS IN GIS

Considering the trends that are currently making an impact in the discipline starts the prediction of what is going to come for spatial problem – solving. There have been many developments in recent days in gathering and analyzing the geographic information and these developments have broadened the horizons of discipline. This has resulted in powerful insights for business, humanitarian organizations, government agencies and the military.

Pervasive shifts have taken place in the way the professional approach the complex issues because of the rise of robust data analytics tools and methods.

Big data is being harnessed by the analytics – focused organizations which are present today. This is being done in order to set goals, prevent fraud, strategize and also, for evaluating the possible outcomes of initiatives and much more.

GIS is implemented in decision – making processes by the organization while tasking experts with obtaining compelling report and visualizations. Furthermore, this allows the stakeholders to make informed decisions.

Let's take an example, there is a manufacturer that can monitor the locations of people and equipment which are involved while producing and shipping the operations in order to determine the need for the expansion of the facilities or streamline its processes.

Over the past several years, along with this, there are growing possibilities for real-time data collection which have resulted in seismic shifts in the GIS field. There are findings from traditional remote sensing sources such as satellite imagery or aerial photography which can be frequently updated.

Furthermore, it can be fleshed out with details from emerging sources including mobile devices and social media. In turn, up-to-date geospatial information tends to speed the responses to natural disasters and humanitarian crises and fuels the services which are based on location such as mobile workforce management, ride - hailing, and proximity marketing.

Open-source mapping has become a crucial part of the way in which the businesses function and individuals get around. There are certain projects such as Open Street Map which have revealed the advantages of involving a large community while submitting spatial data by using the input obtained from GPS devices, field maps and aerial imagery.

Therefore, a map can offer access to local information which is highly detailed; assist multinational corporations that are striving to connect with the customers in a particular area or furthermore, offer intelligence in order to help the workers in monitoring dangerous situations from a distance.

8.3.1. What's next in GIS Technology?

The geospatial perspectives have re-shaped the way in which humans understand the world that surrounds them, and the next wave of GIS software applications as well as techniques is on its way and will come into existence soon. When the organizations tend to embrace the geographic information as a key component in their data strategies, the professionals gain the ability to explore some fascinating questions and also, to deliver meaningful results.

The boundaries of what the human can do with spatial information are expanded because of the development in GIS applications.

There are certain fresh approaches that help in building software programs to be used on the web or mobile devices. These approaches continue to proliferate and furthermore, are facilitated due to the wide availability of GIS APIs and open - source maps.

The apps thus formed can help in educating the government agencies and the public about the history of natural disaster in a specified region, show healthcare professionals the best ways in order to serve the population of any region or teach travelers about nearby landmarks as well as hiking trails.

3-D GIS tools are adopted by most of the developers and because of this, these representations will include layers along with unprecedented levels of detail which can deepen the human's awareness of the relevance of the geographic data.

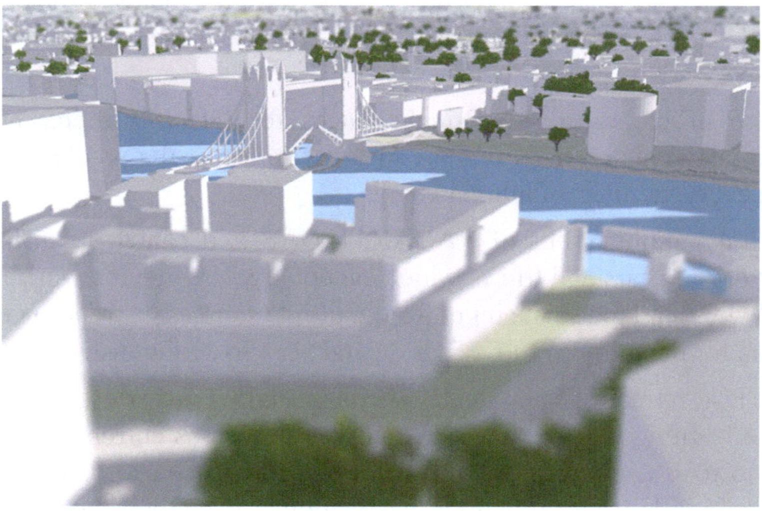

Figure 8.3. 3D London using GD3D base Buildings & trees.

Source: Image by Flickr

A central role is being played by the GIs in the IoT that is the Internet of Things. The reason behind this fact is that a number of organizations use sensors for the collection of information which includes spatial data, from a growing number of devices.

Such kind of observations can further be applied for some of the purposes including improvement in customer engagement in stores and in the identification of the inefficiencies and hazards in the electrical grid. Accurate geo-location allows stakeholders to track the outcomes across supply lines and also, throughout a number of different facilities.

In addition to this, the possibilities of location – based augmented reality have been demonstrated by the retailers and game studios.

These have employed mobile devices in order to superimpose virtual elements in the physical world. A major difference can be made in the richness of an AR app which is done by the geographic precision and up-to-date locational information.

The characteristics of the AR app include being directed to visit a place of business that is recently closed. For instance, this can be devastating for immersion. There can be certain ongoing improvements to the platforms and software that is used by the human so as to participate in such type of experiences. This will possibly make the connections between geographic data and the places that is being visited and that is more involving and responsive.

Some of the most promising chances for GIS are represented by autonomous vehicles. This is done in order to affect the whole industries and the everyday lives of human beings.

Various combinations of RADAR, LiDAR and cameras can be used for self-driving cars and trucks which eventually perceive the world. These technologies tend to use the input from multiple sources in order to generate 3D maps for the purpose of navigation and also, to avoid obstacles.

The fastest route can be calculated by using GIS. Furthermore, it provides machine learning systems having extensive as well as current information related to road conditions and traffic. A big difference can be made while getting a driverless car safely to its destination, especially when there is low visibility because of the weather.

8.4. THE FUTURE OF GIS

GIST that is Geographic Information Science and Technology includes a large number of different ranges of tools and techniques in order to capture analyze and leverage spatial information.

Along with the passing time, there have been advancements in the GIS that are Geographic Information Systems (GIS) which tend to reveal some of the exciting possibilities for space exploration, urban planning, retail and much more. The opportunities are being used by the experts extensively by staying up to date with the emergence of new ways of optimizing and employing GIS technology.

Professionals can capture as well as visualize the valuable geospatial intelligence and this can be done by implementing robust GIS software applications.

This is the reason why it is important for GIST experts to possess the combination of theoretical knowledge along with the hands-on training which comes from attending an online GIST graduate program that is forward - looking.

There are some ways in which the GIS technology can be used in order to meet complex challenges which include a preview of some of the fascinating changes ahead. These ways have been mentioned below.

8.4.1. Open Source GIS Mapping Software Empowers Innovation

Open-source GIS mapping software is the software used for mapping and this software allows the users so that access powerful tools in order to create maps and add elements such as text, images and video.

Specifically, in addition to this, there are initiatives such as Open Street Map or the Open Geospatial Data Project. There can be endless opportunities that can be provided by these applications in order to collect and visualize the spatial information and furthermore, to customize the maps.

The projects are created by the users in order to:

- Trace the availability of natural resources over the period of time;
- Highlight the demographics of a region;
- Demonstrate the way in which the changes in environmental conditions affect the preservation of a historical site; and
- Chart growth of crop and capture any kind of evidence of disease.

There is a growing impact of the flexibility of assembling the maps for diverse purposes in daily life while influencing the progress in the infrastructure of a smart city.

The government agencies and private businesses can deliver services in a more efficient manner and furthermore they can strategize sustainable development along with the extensive locational data. The application of the customized maps can be done to the augmented reality projects which can provide helpful details of a given location to the workers.

In addition to this, it can also be said that the open – source mapping plays a big role in burgeoning the world of navigation systems and automated vehicles. These systems consider a widening range of factors such as traffic conditions updates in real time and individual preferences while setting the routes.

GIS is considered to be important in order to improve the performance in self – driving cars and furthermore, guiding drivers to their destinations comfortably and safely and as quickly as possible.

8.4.2. Fresh Insights from Geospatial Analytics

Over recent years, the most significant shift for businesses is the rise of big data and analytics. Across different industries, there are many organizations that are constantly seeking chances in order to turn an ever - increasing volume of data into a competitive edge.

Among them, a number of organizations have found that GIST has the ability to add up value to these efforts. Geographic information can be applied by the geospatial analytics so as to enable wiser decision - making and agile problem -solving.

Along with GIS data, visualizations can be created by the analytics specialists, meaningful trends can be identified and predictions can be made while generating the reports which flesh out a valuable context for strategy.

The decision - makers gain increased visibility by utilizing the spatial information along with the findings from ERP that is Enterprise Resource Planning systems. Now, they can clearly see the way in which the crucial data about their organizations is attached to particular times and places.

Geospatial analytics tends to clarify the complicated relationships and points out ways in order to streamline the processes. Businesses, as well as government agencies, can:

- Track when devices and equipment with the Internet of Things (IoT) sensors enter or leave a work area;
- Make more informed choices about building or expanding facilities and stores;

- Create apps for retail customers that enrich their in-store experiences (for example, retail beacon technology);
- Find patterns of criminal activity within a region;
- Speed up the logistics and reduce costs by running various routing scenarios; and
- Minimize risks from hazardous location-based events like powerful storms.

8.4.3. Bringing Together GIS and Drones

The UAVs that are Unmanned Aerial vehicles is used by a number of organizations and consumers for an ever-expanding list of applications that can be from entertainment to agriculture. Most of the time, they rely on navigation which is supported by GIS.

For instance, there can be experiments done with making deliveries by drone and they have unlocked the possibilities in order to get food and other products to the customer at an unprecedented speed.

Figure 8.4. Trimble UX5 unmanned aerial mapping system.

Source: Image by Flickr

There are some other companies which are deploying the aircraft in order to collect the information from IoT sensor. This helps in monitoring the industrial sites, volatile weather patterns, hazardous travel conditions or the well – being of crops and sending the real – time updates.

At the same time, the drones have also improved the information which is available for the purpose of mapping and manipulation with GIS tools. A large amount of spatial data comes from the satellite imagery and the readings which are taken by manned aircraft. On the flipped side, it is considered that drones are a cost – effective means for collecting the extensive details on a smaller scale.

Therefore, the UAVs play a major role in making the geographic information more readily available for open – source mapping as well as also, makes it customizable in order to fit the priorities of governments, businesses, independent researchers and hobbyists.

8.4.4. Moving Beyond the Traditional GIS Applications

It's relatively easy to see how GIS technology is transforming the supply chain and asset management, as well as how GIS is essential to cartography and modern maps.

But take a step off the beaten path and consider how GIS might be transforming the virtual world – whether we're talking GIS data visualization or augmented reality and virtual reality.

With the rising acceptance and use of AR and VR in the consumer world, the use cases for leveraging 3D GIS will continue to grow.

As Esri President Jack Dangermond wrote in a recent GeoAwesomeness post, "The geospatial industry is quickly evolving because the capabilities of digital connectedness and collaboration are moving ahead exponentially.

In fact, the five biggest trends in the area of geographic information system (GIS) technology are centered on making data more accessible and creating context to visualize this data in an age when fast, easy access to information is taken for granted."

Esri's City Engine software is one example that Lucy Benton gives that could provide a glimpse of the future of GIS. It allows the production of urban scenes based on real-time, 3D GIS data, so users can plan city architecture, design, and community development.

8.4.5. The GIS AR Opportunity

In the case of the virtual reality and augmented reality the GIS of the future possibly will be well. It is noted that the users are getting more comfortable with the technology due to the continuous growth of AR and VR. Further, it could be GIS capabilities that will be integrated into the applications of the developers as the user satisfaction into technology will encourage the developers.

This is said by Benton. Architects, oil workers and municipal services can also easily use these GIS – enabled AR apps. This statement infers that this technology could be easily used by any person. The orientation of objects underground and to see location these are the major purpose for the use of these technologies.

To serve as the creative mind's playground GIS technology will be used by the game developers in the future to create virtual worlds. There is an expectation to converge the 3D and mobile GIS by the area. Anywhere out in the field immersive experiences will be easily created by the means of 3D spatial information, mobile GIS applications.

Further, the potential of AR visualizations is limitless and there also many other early examples that can be observed in popular games namely Pokémon Go.

8.4.6. Indoor Mapping, UI, AND IoT

To make maps the GIS is used nowadays in the same it used in the above fields, further to develop indoor maps this technology is now being used. As the people will have more need to navigate large, complex buildings in the upcoming years.

So, it is possible that in the future trends in GIS will have indoor mapping more likely skyrocket in popularity. For the emergency responders who act at the points if entry and other critical information when they basically arrive at the scene of the accident, this detailed and more primitive technology of indoor mapping could radically lead to the improved safety.

In this way the technology will turn out is a boon for these emergency responders. It will observe that the future of GIS will come at User Interface (UI) and basically this is one of the biggest changes for developers. Further, it was said by Benton that "These interfaces will make developing work a lot easier because it will feature an easy-to-understand environment that intuitively helps the user to create, add, edit, and remove elements,".

Figure 8.5. The network of IoT.

Source: Image by Pixabay

After that, she further added and wrote, "An introduction of the user interface is a major breakthrough because it will make the maps friendlier to developers as well as everyone who uses them." After this Dangermond said, "Maps are the visual language for understanding the context of data."

The Internet- connected devices that mainly make up the Internet of Things the GIS technology will be mainly beating into the growing network and this was considered as the next important jump in the future of GIS it was basically believed by Dangermond. Further, it is believed that in the real -time GIS practitioners will be able to get even more data, basically by leveraging all the connected devices.

8.5. WHAT'S NEXT FOR GIST?

The future seems richer in possibilities basically for professionals with expertise in spatial technology and problem- solving. As there is a rapid rise in the larger part to the widespread adoption of the smart technology and the IoT it could be predicted that this technology of GIST will likely become even more prevalent aspect of the daily life of the people.

And this widespread of this technology across the large part should be applauded. Further, possibly a growth will be observed in the volume of spatial data, due to the sensory capabilities of sharing spatial- temporal information.

This growth is observed as this sharing spatial- temporal information mainly appears everywhere from industrial equipment to coffeemakers. Moreover, from the spatial software applications, there will a greater demand for functionality and integration.

This demand will mainly put forward by consumers and businesses. Further one of the major advantages which are laid in front of the customers or it can also be called the power of the apps which basically works with GPS and mapping platforms which provide real -time updates.

This power is basically laid in front of the public by ride- sharing services and navigation apps. When harnessed for purposes a growing use for enterprises and government by these types of features. These purposes basically are growing the reach of marketing campaigns and tracking assets.

For the professionals who are familiar with every layer of the GIS technology stack will have a mounting need. The proliferation of spatial data and the market for software means an organization will have a mounting need. And this mounting will be provided to these professionals.

For creating spatial-powered apps to put that information to work there are some of the data resources available. Moreover, it is necessary for the GIS software developers to get familiar with the full scope of the data resources mentioned above. Basically, for collecting, cleaning and formatting spatial information it is important for Data specialists to optimize these processes.

8.6. BRIGHT FUTURE FOR REMOTE SENSING

The progressive societies seek for the prosperity and security and the formation is the foundation of these two. Detailed and reliable information about the Earth's dynamic environment is required for urban planning, energy use optimization, protection from natural hazards, precision farming, climate treaty verification, efficient airline and ship routing and access to natural resources.

Traditionally, it has been observed that this information is basically the domain of academic and professional specialties, aside from the basic weather forecasts. The means for communication results were inefficient, it was difficult to use the tools for turning the data into knowledge and also the raw observations were hard to come by.

Due to the communicating tools for sharing this knowledge, technological advances in sensors, the computing capacity to combine and analyze sensor information all this has changed over the last decades.

Over the next decades, it could be observed that the use of Earth information will be transformed radically due to the emerging result of mass- market consumer's applications. For car navigation mostly the GPS based devices are widely employed.

Everything from education to vacation planning to real- estate hunting is being enabled by the means of the internet maps and virtual worlds (such as Google Earth and Microsoft Virtual Earth). Even in the developing nations the location- based information is provided with an omnipresent distribution platform with the help of mobile phone providers.

Due to the budget cuts and the political decision it is observed that the sophisticated weather satellite system is highly at risk. This weather satellite system is basically included in the Earth information infrastructure in the United States.

Critical climate monitoring measurements are one of the prominent among the potential losses. The third of the U.S. civil satellites will be non-operational by 2010 without decisive action and this mainly reported by the U.S. National Research Council.

8.6.1. Getting the Full Picture

"See and detect from a distance" is the basic function of all the sensing devices although these devices come in a variety of forms, all of them are related to their basic function which is mentioned in the above statement.

The scientific and operational satellites are the major two foundations of the Earth observation system. Supplying everything from the weather observations to regulatory updated global maps of land resources are the major functions of these two scientific and operational satellites.

The resolutions or the perspectives which cannot be obtained from the space are basically provided by the use of aerial systems- aircraft, uninhabited aerial vehicles and even balloons. These aerial systems have basically sent recent revivals.

An important complement is recognized which is namely Ground – based remote sensing systems. One of the examples of the sensing system which is recognized is the Doppler weather radars. At regional and global scales, critical environmental information is provided basically by the low - to moderate resolution imagers.

And these low-to moderate also lays a significant portion of the scientific and operational observations. The weather forecasts shows the "hurricane – from- space" imagery routinely. And this is probably the best- known example. It was basically produced by the means of The National Oceanic and Atmospheric Administration (NOAA).

High- resolution imagers (with a resolution of 1m or better) mostly provide information about human-scale activities such as buildings and roads. The emergence of a small commercial satellite remote sensing industry was possible due to the demand which was led in 1990.

Others were spurred to follow due to the success of these early commercial systems. Including the radar imaging systems with 1- m resolution a number of new commercial imaging satellites will be launched. And this launches will be observed just within the next few years.

Earth's information needs cannot be sufficiently met with the help of imagery alone. The "full picture" about the Earth is basically provided by the imagers as they commonly complement by other sensing techniques.

To determine height profiles of temperatures and humidity, critical parameters for making accurate weather forecasts satellite based spectrometers are employed by NOAA are used. When it is important to "see -through" clouds to the lower atmosphere or grounds the microwaves sensors are routinely called on.

NASA has basically developed more sophisticated sensing techniques. A radar altimeter is used to measure sea surface height to a global average of within 5cm of its actual value. This radar altimeter is employed by The Jason mission (in partnership with France).

To anticipate the arrival of El Nino and the impact of the global warming are assessed by the scientists by the means of this device. To measure the minute changes in the Earth's gravity field is done by the GRACE (Gravity Recovery and Climate Experiment).

GRACE is basically operated by NASA it is a gravity gradient sensor mission (with Germany). The Earth's gravity field is basically used to pinpoint local variations in the planet's density. The slow rise or rebound in the Earth's crust where ice could be readily detected by the GRACE.

10,000 years ago, ice sheets weighed down on the earth were detected by this device of GRACE. Thus, GRACE (Gravity Recovery and Climate Experiment) is acting as a huge remarkable success.

8.6.2. The Next Decade

There is a substantial thrust which is doing things much faster and cheaper and not bigger and better. A number of new techniques are being developed by the military for the purpose of building reconnaissance satellites which can be launched on short notice as needs arise.

It can be possible that the future will expand the use of non – visible portions of the spectrum (specifically infrared and microwave), improve the spatial as well as temporal sampling and also, increase the number of active sensors include LiDAR and RADAR.

The most important thing specifically will be new technologies in order to link the sensors by using the wireless and traditional means into the sensor networks. This in turn, will allow the information that needs to be combined for providing support to the rapid decisions in major and complex situations.

Along with this, the use of the output of one or several sensors can be done in order to trigger the observations from others or also, to reconfigure the other sensors rapidly for optimizing observations of an event.

There is a growing demand, the novel sensor approaches are taken into considerations. Along with this, the novel sensor approaches are also likely to appear.

There is one possibility and that is interactive remote sensing like the farmers tagging their crops genetically in order to enhance the remotely detectable spectral signature for crop distress or optimal harvesting.

There are major efforts put by the policies and these are underway for guiding this future. In the year 2003, a global initiative was spearheaded by NOAA which was referred to as GEOSS which means Global Earth Observation System of Systems. Through this, a loose coalition was made of 66 nations who aimed at coordinating much of the Earth -observing activities of the world.

The main focus of GEOSS was on using the resources that are present in more effectively and this is surely a worthy goal. A report was present by the National Research Council (in the US).

This report referred to earlier laid out of a decadal plan specifically for Earth observation from space. This included a set of 17 new missions which are to be launched during the upcoming decade.

There are certainly competing forces which are of limited government resources, increased business demand and advancing the technological capability, which will play out during this time period.

The consumer needs are growing rapidly which will eventually introduce a new as well as unpredictable factor. The future thus seems to be bright, but this will definitely not be happening on its own. The communities of professionals are required to work diligently in order to ensure its success.

8.7. NEW TECHNOLOGY TRENDS

The themes are apparent in the development of technology for Earth remote sensing which will likely continue for decades in the upcoming future. There are certain key optical and radiofrequency technologies for the future systems which have been described. If in case, there is one underlying technology area that governs the rate of progress in all these areas, it seems to be materials.

The new materials tend to enable shrinking circuit sizes. Furthermore, they allow the lower power operation for 3-D packaging. The new materials extend the spectral range of the performance of detectors.

They tend to provide strength along with the low mass for large structures. They enable high power, compact optical systems and also more efficient transmitters.

A number of these trends are apparent in the development of technology and so are being funded today by the ESTO that is Earth Science Technology Office for the Earth Science enterprise of NASA. Many of these trends have been mentioned below.

8.7.1. Miniaturization of Electronics

The miniaturization of electronics has been particularly referred to as digital electronics and their progress is much known. In this, size decreases and the wafer size increases and due to this, a large number of gates can get fit on a single integrated circuit. This is a trend that will continue for many upcoming years and a number of new trends will become apparent in the next upcoming decade.

The 3-dimensional chip making is used and this will become widespread because of the fact that it solves many cooling and connection problems. The overall packaging sizes will reduce dramatically as there is a decrease in the order of magnitude in volume and mass.

In analog electronics, there is a similar trend that is occurring. Let's take an example; the MMICs that are monolithic microwave integrated circuits are used and furthermore, this is enabling significant decreases in the size as well as in the mass of radiofrequency circuits. MMICs are still relatively new and also, the higher integration levels will be followed in the upcoming future.

Along with this, there have also been decreases in the size of analog electronics, and due to this a number of analog systems will get evolve to fully digital systems along with the antennas that are connected directly to analog – to – digital converters.

As there is an increase in the sensitivity and in format size, the integration of the detectors will be done along with all of the ancillary electronics into a single 3-D chip. This chip could replace all of the electronics of one of the instruments that are being used today.

Furthermore, considering the future, it can be said that the electronics will become simply a part of the structure having no separate packaging. This will happen because of the fact that the membrane and 3-D electronics are attached directly to structural members.

One more reason can be that the electronic components actually become load- bearing elements themselves. These trends have been furthered because the ESTO projects develop high - frequency MMICs, precipitation radar and a photovoltaic long-wave infrared detector.

There have been astounding increases in processing speed along with the decrease in the size of electronics for both programmable as well as general -purpose processors.

Till this time, the technologies which are being used are fully exploited. There are radical new technologies including quantum and biological computing which will be realized.

There has been a steady progression in the storage density which will continue further. For all the practical purposes, the final result will be such that the computing power may become unlimited on orbit within the sensor itself. The development of holographic memory is funded by ESTO which is a step in such a direction.

8.7.2. Large, Lightweight Structures

There are measurements that have higher spatial resolution and these measurements will require larger apertures. There will be progress in the technology used for large deployable antennas and mirrors steadily.

Today in this scenario, the approaches to large deployable mesh and membrane antennas are being pursued. This usage will be extended to larger and larger systems of meters with lower and lower mass density.

There will be an advancement in the deployable mirror and adaptive optics technology in order to enable low-mass as well as multi-meter optical systems. The work of ESTO on a 6 -meter inflatable radar antenna and large, sparse aperture radiometers are considered to be examples of this progress.

8.7.3. Increased Power Active Sensing

A number of LEO active sensors have been deployed successfully. Still, there are a number of challenges that remain while increasing the power and efficiency of lasers and radars. It is important to have increased power levels in order to extend the radar and laser measurements to GEO and also, to challenge the DIAL and tropospheric wind measurement applications considering the LEO.

There are materials such as new laser and non – linear, frequency – shifting materials which are possibly considered to be a fundamental part in order to achieve these long-term goals. These are increasing the power levels and efficiencies in solid – state radiofrequency amplifiers. There have funds provided by ESTO to a Laser Risk Reduction Program which tend to address some of these laser challenges.

8.7.4. Compact Optics

By using the smaller instruments, there can be increased coverage through multiple copies of sensors. There have been a number of innovative approaches to refractive and reflective optics and they are under development. There can be a reduction in the overall size and mass of optical instruments which will be allowed because of the Compact Fourier transform and grating spectrometers.

The extension of these size benefits far into the infrared will be done due to the efficient new coolers. The ESTO has provided funds for the technology for an infrared sounder which tends to use compact refractive optics as well as work on an optical cryo cooler.

8.7.5. Frequency Flexibility

In today's era, the selection of the wavelengths and their corresponding frequencies are done in a careful manner and are sparse because of the fact that the limited resources constrain system configurations. In the upcoming future, the hyper-spectral and tunable systems will be commonplace. Moreover, this will allow much greater flexibility in the spectral domain.

A number of observations are allowed by the tunable lasers by using the same instrument and adaptation of wavelengths in order to provide the most accurate and précised measurements.

Hyper-spectral systems will further allow flexibility by performing the on-board frequency selection, aggregation, and also, combination. Wider bandwidth RF systems are allowed due to the immediate digital conversion of signals from antennas.

Detectors will become multispectral themselves, and in some cases, it eliminates the need for spectrometers. The funds have been provided by ESTO to a new hyper-spectral infrared imaging detector. This detector tends to image four infrared bands in the 3 to 15- micron range on one detector.

8.8. FUTURE SENSORS: NANO-SATS AND WISP-SATS

Considering the next several decades, the technology trends have resulted in a very different image of the set of remote sensing satellite from that image which it there today. The highly connected nature of many affordable as well as very capable, flexible satellites have resulted in sensor web.

Earlier, there have been a number of papers in IGARSS sessions that have discussed the sensor web, and along with this, the nature of some of the individual future sensors.

The technology trends have caused vast increases in electrical and mass efficiency which have discussed above and this will result in a more heterogeneous set of sensor systems.

The entire size of the sensor system will be defined by the aperture which is required for the larger – aperture systems and probably, it can also be required by the solar power collection system for the small satellites.

These sensors have a range of sizes that will far exceed in comparison to today along with some very large systems having large apertures and the smallest systems no bigger when compared with a bread box. Such types of systems may appear to be flying antennas or flying solar arrays with no traditional spacecraft bus from a distance. Everything will be disappeared but those structures where size is dictated by physics will remain.

There will be more even distribution of sizes from centimeters to meters instead of the distribution of satellite sizes which is today which cluster together in the several meter size range. This depends on the application.

The former are the micro- and nano-sats and the latter. They have their very large but lightweight structures which can be referred to as wisp-sats because of their delicate as well as insubstantial appearance.

There are certain scatter plots of points that represent the Earth science satellites in today's time. These may include Aqua, ACRIMSAT, TOMS-EP, Terra, SOURCE, ICESat, GRACE, Landsat 7, Jason-1, EO-1, and RADARSAT. The shape of oval represents those areas which will become increasingly populated in the upcoming future.

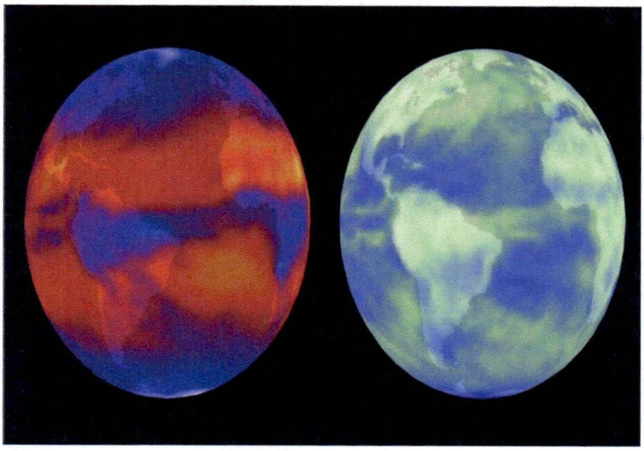

Figure 8.6. Terra/CERES views the world in outgoing long wave radiation (left) and reflected solar radiation (right).

Source: Image by NASA

There has been a progression to large numbers of small satellites which eventually will depend on an increase in launch opportunities and also, the decrease in the launch cost. When there will be progression to very large satellites then this will depend on the progress of the enabling large structure technology.

8.9. CONCLUSION

There is rapid growth taking place in the geospatial science and engineering field and therefore, it is expected to get advanced significantly in the upcoming years. Primarily, the developments are fueled by a large amount of geospatial data and the availability of big data methods. This data is acquired by ever-powerful sensors, both professionally as well as the crowd – sensed. The main question that arises here is that 'how geospatial data can harness the data science evolution that is been ongoing as it has been cleared that the traditional processing cannot keep up along with the volume of data coming from the satellites, smart devices, UAS, driverless vehicles, etc. and so, is ever - increasing volume.

The key element of this process is exploiting the potential of crowd - sensed data because of the fact that it provides for sustained observation of the environment while offering both currencies as well as accuracy.

The conclusion that can be drawn from the above chapter is that the integration and miniaturization are increasing which will mean that the spacecraft will take on the shape as well as the general characteristics of their largest physical constraint.

REFERENCES

1. Dangermond, J., 2013. *GIS: Transforming Our World*. [online] Esri. Available at: <https://www.esri.com/about/newsroom/arcnews/gis-transforming-our-world/> [Accessed 22 April 2020].
2. Gail, W., 2007. *Bright Future For Remote Sensing?*. [online] Spie.org. Available at: <https://spie.org/news/spie-professional-magazine-archive/2007-july/sp-bright-future-for-remote-sensing?SSO=1> [Accessed 22 April 2020].
3. Geoawesomeness. 2019. *The Future Of GIS - Geoawesomeness*. [online] Available at: <https://geoawesomeness.com/the-future-of-gis/> [Accessed 22 April 2020].
4. Hartley, J., n.d. *Earth Remote Sensing Technologies In The Twenty-First Century*. [ebook] p.2. Available at: <https://esto.nasa.gov/conferences/igarss03/papers/TU091640_hartley.pdf> [Accessed 22 April 2020].
5. Sajjad, H. and Kumar, P., 2019. *Chapter 16 Future Challenges And Perspective Of Remote Sensing Technology*. [ebook] p.2. Available at: <https://www.researchgate.net/publication/329178300_Future_Challenges_and_Perspective_of_Remote_Sensing_Technology_Potential_and_Future_Trends> [Accessed 22 April 2020].
6. Toth, C., n.d. *THE FUTURE OF REMOTE SENSING: HARNESSING THE DATA REVOLUTION*. [ebook] p.1. Available at: <http://ppct.caicyt.gov.ar/index.php/geoacta/article/download/12533/45454575757740> [Accessed 22 April 2020].
7. USC GIS Online. 2019. *The Future Of GIS*. [online] Available at: <https://gis.usc.edu/blog/the-future-of-gis/> [Accessed 22 April 2020].
8. USC GIS Online. 2019. *Top GIS Technologies In 2018 & Beyond*. [online] Available at: <https://gis.usc.edu/blog/top-gis-technologies/> [Accessed 22 April 2020].

INDEX

Printed in the United States
By Bookmasters